TRANSPORTATION NETWORK MODELING AND CALIBRATION

TRANSPORTATION NETWORK MODELING AND CALIBRATION

MANSOUREH JEIHANI AND
ANAM ARDESHIRI

MOMENTUM PRESS, LLC, NEW YORK

Transportation Network Modeling and Calibration

Copyright © Momentum Press®, LLC, 2017.

First published by Momentum Press®, LLC
222 East 46th Street, New York, NY 10017
www.momentumpress.net

ISBN-13: 978-1-60650-893-0 (print)
ISBN-13: 978-1-60650-894-7 (e-book)

Momentum Press Transportation Engineering Collection

Cover and interior design by Exeter Premedia Services Private Ltd., Chennai, India

10 9 8 7 6 5 4 3 2 1

Printed in the United States of America

To the love of our lives

Mario and Annika

and

Maryam and Nojan

ABSTRACT

Transportation scientists employ modeling and simulation techniques to capture the complexities of transportation systems and develop and assess solutions to alleviate existing and future transportation-related problems. This book introduces transportation engineering students and junior engineers to the concept of transportation network modeling, network coding, model calibration and validation, and model evaluation. Transportation networks are built via links and nodes to replicate roadways and their junctions. Transportation network elements, demand and supply models, and equilibrium concepts are reviewed in this book.

Travel demand models are sensitive to demographic changes and can explain and forecast how a new transportation supply system leads to a new transportation demand pattern. Discrete choice analysis theory and its application in travel demand models are discussed in this book. Most of the traffic assignment models introduced in the literature are covered. The widely utilized traditional four-step modeling approach, which is still the first choice in many American metropolitan planning agencies, is elaborated. This book also describes how demand models evolved from trip-based to the newer generation of activity-based and agent-based to overcome some of the shortcomings of the four-step approach and improve models' prediction power.

One major focus of this book is model calibration, error checking, and fine-tuning. Calibration techniques in planning and simulation, common challenges, and steps to minimize calibration errors and improve model validity are introduced.

KEYWORDS

activity-based models, agent-based models, calibration, four-step models, network modeling, simulation, system optimal, traffic assignment, travel demand analysis, travel forecasting, user equilibrium, validation

CONTENTS

LIST OF FIGURES

LIST OF TABLES

ABBREVIATION LIST

A	Attraction
AADT	Average annual daily traffic
ABM	Agent-based modeling
ATIS	Advanced traveler information systems
BMC	Baltimore metropolitan council
BPR	Bureau of public roads
CA	Cellular automation
CART	Classification and regression tree
CATS	Chicago area transportation study
CMS	Changeable message signs
CBD	Central business district
CTPP	Census transportation planning package
D	Destination
DMS	Dynamic message signs
DS	Driving simulator
DTA	Dynamic traffic assignment
EB	Eastbound
EBT	Eastbound left
EBT	Eastbound through
FFS	Free-flow speed
FFT	Free-flow time
HBO	Home-based other
HBS	Home-based school
HBSh	Home-based shop
HBW	Home-based work
HCM	Highway capacity manual

HOV	High occupancy vehicles
HPMS	Highway performance monitoring system
IPF	Iterative proportional fitting
ITS	Intelligent transportation systems
LOS	Level of service
MLE	Maximum likelihood estimation
MNL	Multi nominal logit
MOE	Measure of effectiveness
MPO	Metropolitan planning organization
NB	Northbound
NBT	Northbound left
NBT	Northbound through
O	Origin
OD	Origin–destination
P	Production
PA	Production attraction
PHS	Peak hour spreading
PUMA	Public use microdata area
PUMS	Public use microdata sample
RTOR	Right turn on red
RP	Revealed preference
SO	System optimal
SP	Stated preference
STA	Static traffic assignment
SUE	Stochastic user equilibrium
TAZ	Traffic analysis zone
TRANSIMS	Transportation analysis and simulation systems
TDSP	Time-dependent shortest path
TMC	Transportation/traffic management center
UE	User equilibrium
VHT	Vehicle hours traveled
VMT	Vehicle miles Traveled
VOC	Volume over capacity
VOT	Value of time
VPH	Vehicle per hour

VPHPL	Vehicle per hour per lane
WB	Westbound
WBT	Westbound left
WBT	Westbound through

PREFACE

Transportation science is established to illustrate and evaluate the quality of people and goods' movement along transportation infrastructure. A high demand for transportation professionals is anticipated in the near future in both developed and developing countries, especially in the field of intelligent transportation systems (ITS) and innovative solutions for urban transportation challenges.

This book intends to address the gap between the material in the traditional transportation engineering textbooks and what a future transportation engineer or modeler is expected to perform. With the rapid change in software development and data collection technologies, although the traditional materials offer the core knowledge of transportation science, they do not provide sufficient guidance for practitioners to develop, calibrate, and validate transportation network models.

This book is a result of over 12 years of experience in transportation network modeling. We have approached the subject from a modeling exercise, discussing theories, data, model estimation, calibration, and validation. Our aim in writing this book was to create a textbook for transportation students at both the undergraduate and graduate level, as well as a reference volume for junior practitioners. Step-by-step examples for a four-step model and an activity-based model would help students learn the concepts thoroughly and prepare them for jobs in transportation planning and modeling fields.

Chapter 1 is the introduction to a transportation network and its components, along with transportation demand and supply, and equilibrium concept. Roadway characteristics—such as speed limit, number of lanes, lane width, direction(s) of flow, heavy vehicle percentage, and intersection control type—are key elements being coded when developing a mathematical model. Model aggregation and zoning requires additional components, i.e. dummy nodes and links, to load a network with traffic data. Traveler's behavior, which is an introduction to transportation demand modeling, is explained in Chapter 2. Chapter 3 covers various traffic assignment models that include how the estimated trips generated by each vehicle class are allocated on the roadway network. Travel demand modeling approaches, four-step models, activity-based models, and agent-based models are introduced in Chapter 4. The focus of Chapter 4 is on the most commonly used modeling approach, four-step models, due to their easier data accessibility than the more advanced methods. Real-time systems and demand management using real-time traffic information is explained in Chapter 5. Calibration and validation techniques in transportation demand models as well as in traffic simulation models, along with real-world examples, are explained in Chapter 6. Model calibration is an iterative process that adjusts a model's parameters to reproduce calibration parameters that are identified and measured prior to the modeling effort. A model should be validated with a new dataset that was not included in the calibration process to gauge the model's performance and reliability. Chapter 7 concludes the book with a quick review and summary of the highlights in transportation network modeling.

We would like to thank Dr. Brian J. Katz, Dr. Ardeshir Faghri, and David B. Roden for their invaluable comments and feedbacks; Nancy Jackson for editing the book; and Zohreh Rashidi Moghaddam for providing the figures and tables. Special thanks to my PhD advisor, Dr. Antoine G. Hobeika, the base of my transportation planning knowledge stems from him.

CHAPTER 1

INTRODUCTION

Transportation is the means to reach a destination. It facilitates people and goods being where they need to be at a certain time. Transportation science is established to illustrate and evaluate the quality of people and goods' movement along transportation infrastructure of all shapes, including ground, rail, waterway, air, and pipeline. Similarly, many other facts evolve in the rapidly changing world, and travel patterns and travelers' behavior are constantly transforming to adapt to today's lifestyle and respond to new transportation needs. Transportation models are capable of calculating traffic consequences, including travel delays, queue backups, environmental impacts, energy consumption, and crash rate.

The first step of modeling transportation is to develop a transportation network of the study area as a setting for people to travel. A transportation network comprises several items, with each representing one real-world element in our daily travel life. Network links denote urban streets, while nodes represent street junctions. Vehicles and travelers are moving objects in a transportation network in the direction(s) defined for the links. A transportation network shows the origin and destination of traffic, from which node vehicles can enter and exit the network. All feasible and allowed movements or parking restrictions (by time of day) are reflected in the transportation model. The notion of a supernetwork indicates the basic transportation network augmented with dummy links. This concept is beneficial when combining travel demand problems in different stages and solving a joint equilibrium problem. In summary, what is necessary in network modeling may comprise the following:

- Direction of flow
- Capacity of roadway per direction (maximum flow rate)
- Traffic control system at junctions or intersections
- Traffic volumes
- Pedestrian and bike activities

- Heavy vehicles
- Traveler awareness system (about travel information)
- Traffic mix, such as car, bus, and bicycle
- Link speed or travel time
- Link interaction (travel time versus volume of the link)

This chapter begins with a review of transportation demand, supply, and equilibrium concepts. Later, it explains the key components of transportation networks and their roles in the models.

1.1 TRANSPORTATION DEMAND

The public definition of transportation demand is the need for transportation services that generate movements of people and freight. Demand for transportation is the willingness to pay for transportation services and how this willingness changes when the price changes. The economic definition of transportation demand is the relationship between the quantity of transportation services demanded (consumed) and the price people are willing to pay for it. This relationship is shown with a curve. The demand curve is the downward slope as presented in Figure 1.1, which presents the quantity demanded versus price. When the price of transportation service increases, demand decreases. Examples of quantity demanded are vehicle per hour, number of passengers per day, and tons per day (for freight).

1.2 TRANSPORTATION SUPPLY

The capacity of infrastructure and transportation mode over a geographically defined transportation system and a specific period of time is called transportation supply. The economic definition of transportation supply is the relationship between the quantity of transportation services supplied

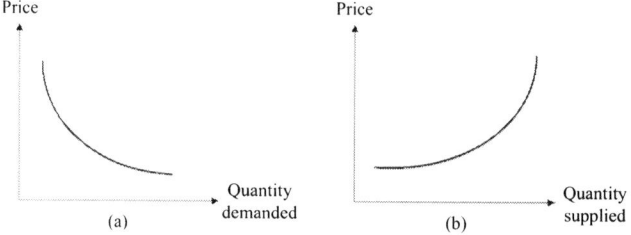

Figure 1.1. (a) Demand curve (b) Supply curve.

(offered) and the price charged for it. The supply curve is an upward slope as presented in Figure 1.1, which presents the quantity demanded versus price. Examples of the quantity supplied are vehicle per hour, number of passengers per day, and tons per day (for freight).

1.3   EQUILIBRIUM

Equilibrium is defined as the price in which the quantity demanded and the quantity supplied is equal. Figure 1.2 presents the equilibrium (P*, Q*). In other words, equilibrium is the point where the price of transportation is just right, so that the quantity demanded is entirely supplied. If the price is higher than P*, then the quantity supplied is more than the quantity demanded, resulting in a surplus. If the price is lower than P*, the quantity demanded is more than the quantity supplied, resulting in a deficit.

When demand increases (decreases), the demand curve shifts up (down), and so, the equilibrium price and quantity both increase (decrease), as presented in Figure 1.3. It means that more (less) transportation services are purchased at a higher (lower) price. As presented in Figure 1.3, when supply increases (decreases), the supply curve shifts to the right (left),

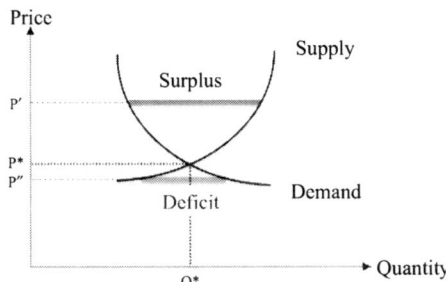

Figure 1.2. Equilibrium, surplus, and deficit.

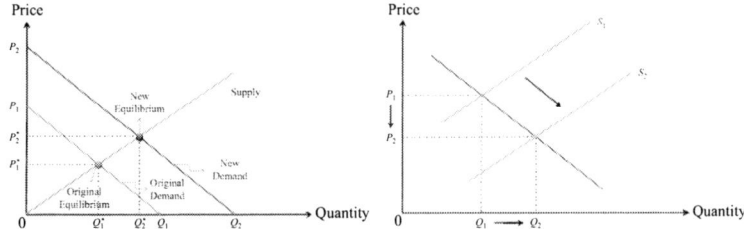

Figure 1.3. Change in equilibrium when (a) demand increases (b) supply increases.

reducing (increasing) the equilibrium price while the equilibrium quantity increases (decreases). When both the demand and supply increase, the equilibrium quantity increases, but the price might increase, decrease, or stay the same, depending on the amount of increase in demand and supply.

1.4   TRANSPORTATION DEMAND AND SUPPLY MODELS

Transportation modeling comprises of modeling models, both supply and demand sides of transportation. While the supply side is less dynamic and more predictable, modeling the transportation demand can be very challenging considering the interaction between drivers and reaction to the change in other factors, such as weather, work zone, school zone, and incident. Transportation demand models formulate and predict transportation demand. The demand is usually described by origin, destination, and mode, time, and path of travel. The demand may be estimated by the trip region (e.g., home-based) or by special purposes (e.g., work trips) or destinations (e.g., airport).

Transportation supply models predict the performance of transportation systems, given the demand. They utilize the traffic flow theory and network flow theory to simulate the transportation supply system. The predicted performance includes origin destination (OD) travel times and costs by mode of travel and by time of day, and estimated link volume.

1.5   TRANSPORTATION NETWORK COMPONENTS

The supply side of a transportation system can be represented by a network, which includes a set of points and a set of lines that connects the points. A node (vertex or point) represents an intersection, a transit stop, or an activity center (office, house, shopping center, etc.), and a link (arc or edge) represents a street, highway, transit route, and so on. A transportation network is usually a directed network, representing the direction of flow.

Each link in the network is associated with some impedance, usually travel time of the link. The link travel time is the time it takes to traverse a link and depends on the link flow. When there are no other vehicles on the road (free flow), the driver can drive at any desired speed within the posted speed limits, but when other vehicles exist on the road, travel time increases due to interaction with those other vehicles.

Transportation networks can be created using geographic information systems (GIS), which are very accurate. Typically, all freeways, expressways, principal arterials, minor arterials, and collector routes are included in a transportation network. Local roads are usually excluded in aggregated models.

Transit networks include transit routes, some of which share highway links, while others are on their own right-of-way. Transit networks are usually more complex than highway networks because of the multiple modes involved and the need to consider operating frequencies and schedules.

1.5.1 NODE

On a road network, a node represents an intersection, the end of a road, where the road branches out, or a change in the number of permanent lanes. It can also represent a transit stop on a transit network. Nodes can be origin, destination, or transshipment nodes. When creating a node in the network, allowed and blocked movements, particularly left turn movements, should be coded accurately. Figure 1.4 depicts a four-leg intersection with all turning movements allowed coded in a simulation tool. The highlighted connector shows double left lanes for eastbound left movement. A node is a point in two dimensions, the coordinates. Nodes are recorded in a table with Node ID, Easting (X-coordinate), and Northing (Y-coordinate).

Figure 1.4. A simulated node with all potential turning movements.

Table 1.1. A sample of a node table

Node ID	X coordinate	Y coordinate	Centroid	Zone	Notes
2001	150	500	1	12	
2002	300	500	0	12	
2003	100	100	0	3	

Other information such as zone number, node type (centroid or real), and so on can be added to the table (see Table 1.1).

1.5.2 LINK

A link connects two nodes and represents a roadway or a transit route. Link Table includes link ID, Link Type (urban, suburban, rural), functional class (freeway, arterial, etc.), number of lanes, and so on (see Table 1.2). Figure 1.5 presents links that connect nodes A to E. Five links are bi-directional and four links are unidirectional. The numbers on the links represent (free flow) travel times in minutes.

Some important link characteristics that need to be coded when modeling a transportation networks are as follows:

- Length: The length of the link needs to be accurately coded to represent the real-world roadway (transit route) length.
- Free-flow speed (FFS): It is usually higher than the speed limit (e.g., 20 percent higher) to represent the speed of vehicles when the link is almost empty and the driver can drive freely.
- Road class (or facility type): Roads are typically classified into freeway, expressway, major arterial, minor arterials, local, and centroid connectors. A number is allocated to each road class (Table 1.2).
- Capacity: The maximum number of vehicles per hour per lane that can be accommodated on the road. Usually, capacity is the same for each road class.
- Direction: To specify whether the link is bidirectional or one-way and the direction of traffic in the latter case.
- Travel time: Before the network is loaded and there is no vehicle on the network, travel time is free-flow time (FFT) and is calculated as length divided by FFS. However, when the network is loaded, travel time is calculated by link performance function or simulation. In link performance function, it is assumed that travel time is a function of volume and capacity of the link. In simulation, travel

Table 1.2. A sample of a link table

Link ID	From node	To node	Length	Class	Lanes	FFS	Capacity per lane	Dir	Time (min.)	Counts	Notes
103	2001	2002	2.2	1	6	65	1100	0	2.03	4000	
104	2002	2003	1.5	1	6	65	1200	0	1.38	4200	
251	2003	1025	.2	5	2	30	700	1	0.4	1000	
252	1025	1026	.5	3	3	45	800	0	0.67	1500	

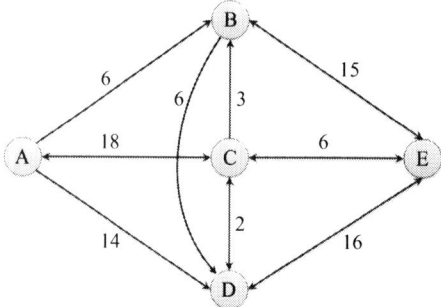

Figure 1.5. A road network with five nodes and nine links.

time experienced by vehicles is recorded and averaged for each link. The link performance function and simulation will be covered in Chapter 5.
• Number of lanes: Number of lanes of the roadway.
• Traffic count: It is the current traffic volume on the link. This information is not available for all links. This information is used for calibration.

The preceding attributes of the links are important in network modeling and must be accurately coded. Failing to correctly code them results in inaccurate models. Some other link characteristics such as road name may be included. When coding a 24-hour model, time-of-day restrictions of left turn movements, reversible lanes, and parking restrictions should be considered to reflect an accurate number of operating lanes at any time.

1.5.3 ZONE

In the transportation planning process, the study area is usually divided into traffic analysis zones (TAZs). The size of each TAZ varies from a city block to a neighborhood. Traditionally, TAZs are defined in such a way that a specific activity is in the majority. A TAZ can be residential, commercial, industrial, and so on. The boundaries of TAZs are usually manmade, such as roads, or nature-made, such as a river. The number of TAZs in a study area may vary from several dozens to several thousands. A common practice is to have TAZs no more than 3,000 people.

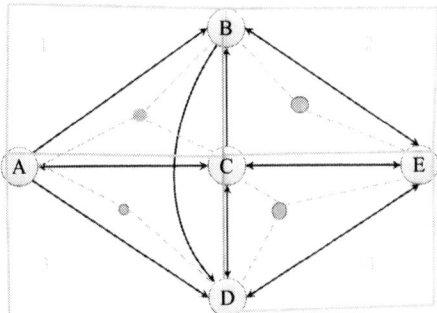

Figure 1.6. A road network with four zones.

TAZs are usually defined based on the following criteria:

- Attain homogeneous socioeconomic characteristics for each zone
- Minimize the number of intra-zonal trips
- Recognize physical, jurisdictional, political, and historical boundaries
- Define TAZ boundaries based on census zones
- Specify TAZs in which the number of households, population, area, or number of trips produced and attracted are almost the same
- Number of households, population, area, or trips generated or attracted should be nearly equal in each TAZ
- Zonal boundaries are preferably based on census zones boundaries

Other than internal zones, a few external zones may be defined to capture the effect of external trips to the study area, which may include external-to-external trips, external-to-internal trips, and vice versa. External TAZs are normally larger and include less detailed transportation networks. Figure 1.6 presents four TAZs shown in green polygons; some of their boundaries are roads. TAZ numbers are shown in green.

1.5.4 CENTROID AND CENTROID CONNECTORS

In aggregate modeling, each zone is represented by its center of activity, called zone centroid. It is assumed that all vehicles originated from and are destined to the centroid of the zone. The center of gravity could be the geometrical center of gravity, center of gravity of the trips' origins and destinations, center of gravity of weighted network nodes, or the node with the highest accessibility index in the zone.

Zone centroid is connected to the roads via dummy (imaginary) links called centroid connectors. It is assumed that the centroid connectors have high capacity so that there is no traffic congestion on those imaginary links. Centroid connectors represent local streets that connect homes and businesses to higher-level roads, such as collectors and arterials. They are the first and last leg of each trip and represent access from neighborhoods and businesses to the roadway and transit systems.

Ideally, centroid connectors should be attached to a road functional class higher than local roads (such as collectors or minor arterials) at locations where local streets connect to them, but they may also occur where commercial or residential activity locations directly access the road network. There are practical limitations to the number of centroid connectors.

In general, centroid connectors are attached at mid-block or at inter-sections. Connection at mid-block is preferable because it yields more detailed smoothing of traffic assignment results and a more efficient level of service analysis at intersections. Specially, if the turning movements at intersections are important in the study, connecting centroid connectors to the intersection is not plausible. Some guidelines for centroid connector placement are as follows:

- Attach connectors to the roads at or near the TAZ boundary, but do not cross the boundary.
- Connectors should not cross natural or man-made features that prohibit such crossing, such as rivers or railroads.
- Do not connect centroids directly to interstates, freeways, or ramps. Attach them to the lowest functional classification possible (e.g., collectors, minor arterials).
- Limit the number of centroid connectors to a practical amount (one to three connectors) per TAZ.

1.6   THE CONCEPT OF MODELING

Models are simplifications of reality that try to show how a complex system works. Like any model, transportation models are a mathematical formula based on a set of assumptions, some of which are unrealistic, and the model represents the current situation of the system. A transpor-tation model is developed for a geographical area, such as a corridor, city, county, and state.

Transportation network modeling discovers how much flow goes from each origin to each destination while minimizing cost. An assignment model assigns each vehicle, person, or a group of them to a route from their origin to their destination. Transportation models can be behavioral or aggregate. Transportation network modeling is a mixture of art and science; the supply side modeling is science and the demand side is art.

Transportation modeling attempts to answer the following types of questions:

1. What is the utilization of transportation facilities?
2. How does users' utilization of the transportation system change over time?
3. How do users respond to changes in the transportation system?
4. How can planners improve a transportation system to accommodate future utilization of the system?

1.7 PLANNING VERSUS REAL TIME

Transportation network modeling can be for real-time or planning applications. In real-time, the models assign traffic based on real-time data obtained by traffic monitoring systems, including loop detectors, wireless sensors, radars, and video cameras. The model is updated every so often, for example, every five minutes. Using historical data and data mining techniques, real-time systems can perform traffic prediction for near-future terms, which can be extremely beneficial in congestion management and active transportation demand management.

In planning applications, a base year is selected and the information for the base year (such as traffic counts, population, employment, and land use) is utilized to make the base model, which represents the current conditions. The base year is then calibrated and validated with the real data. The calibrated or validated base model is then used to build the future models and forecast future traffic conditions and test improvement scenarios.

Because traffic congestion peaks during some known hours of morning and evening, it is a common practice to focus on AM and PM peak periods in modeling a transportation network. Midday (MD) models, as well as weekend and special events, are also commonly analyzed individually for a network's performance evaluation of nonpeak periods.

BIBLIOGRAPHY

deOrtúzar, J.D., and L.G. Willumsen. 2001. *Modelling Transport.* Chichester, United Kingdom: John Wiley & Sons, LTD.

Friedrich, M., and M. Galster. 2009. "Methods for Generating Connectors in Transport Planning Models." *Transportation Research Record, Journal of Transportation Research Board* 2132, pp. 133–42.

Mankiw, N.G. 2015. *Principles of Microeconomics*. 7th ed. Boston, MA: Cengage Learning.

Meyer, M., and E. Miller. 2001. *Urban Transportation Planning: A Decision-Oriented Approach*. 2nd ed. New York, NY: McGraw-Hill.

Sheffi, Y. 1985. *Urban Transportation Networks: Equilibrium Analysis with Mathematical Programming Methods*. Upper Saddle River, NJ: Prentice-Hall.

CHAPTER 2

TRAVELERS' BEHAVIOR

Transportation demand models are the means to explain and forecast human behavior. These models strengthen our knowledge about travelers' opinions and attitudes toward transportation supply systems and their reactions to any change to the system. Knowing more about how travelers make their decisions helps transportation planners to better manage their transportation problems and respond to the variability in demand. This chapter discusses the human factors in transportation and reviews modeling advances in this arena.

Our behavior is driven by a choice we have made when there is a choice set available to us. To better understand urban travelers' behavior, it is beneficial to dig into the variety of applications that play a key role in travelers' choices. These applications begin with: whether or not to make a trip, where to go, what time to start a trip, what mode of travel to choose, and in the case of driving, which route to take, what speed to drive, and so forth. Therefore, a transportation modeler is expected to determine all trip elements for a given trip origin to be able to estimate the impact of any individual trip on the transportation network. On the other hand, transportation impacts many other choices in our lives, such as choosing our residence. It determines the distance from origin to destination for any trip purpose, the transportation mode options available to us, traffic status, and network reliability, among others.

Human choices form the skeleton of transportation demand models, and therefore, understanding travelers' decision-making process substantially assists transportation modelers and engineers to analyze existing transportation problems and predict travelers' behavior upon any changes in the transportation system. Following a general description of choice theories, this chapter reviews the major uses of choice models in transportation modeling. Choice models in the transportation modeling framework account for behavioral aspects of this framework and enable modelers to better explain travelers' behavior. They are usually built in a disaggregate level demonstrating users' heterogeneity and the complexity of human choices.

2.1 DISCRETE CHOICE ANALYSIS

A traditional view of travelers' choice indicates that people make decisions that maximize their utility with the information available to them at the time of decision. Many factors that affect the utility functions are considered as random variables. Reasons such as heterogeneity in traveler preferences, calculation errors, unobserved attributes, and insensitivity to small changes in attributes support the notion that utility should be treated as a random function.

Utility functions are generally comprised of deterministic and stochastic segments. In the deterministic segment, the output is fully determined by the parameter values and the initial conditions. A stochastic segment includes some inherent randomness. The categorization of different discrete choice models is based on the mathematical model structure for the deterministic segment. Logit structures (such as multinomial logit, nested logit, C-logit, and path-size logit), generalized extreme value (such as generalized nested logit), and non-generalized extreme value (such as multinomial probit, logit kernel) were among the commonly tested structures in travel demand choice-related analysis. The selection of suitable structure, however, depends on the size, distribution, number of viable alternatives, and other features of the transportation network and study data.

The "utility" indicates the value of each choice to each traveler. For a route choice example, route r is chosen by traveler t if the utility of that route perceived by that individual (U_{rt}) is greater than the utility of all other routes in the traveler's choice set. This concept implies that utilities of different route options can be ranked, wherein the one with the highest value is ultimately selected. Therefore, what is important to determine the optimal choice is the difference in the utility between any two choices, not the absolute values of utility functions.

Because full knowledge of a traveler's decision-making process and his or her perceptions toward alternatives are not available to analysts, utility functions are subject to uncertainties. To tackle this uncertainty, utility is decomposed into two components of deterministic (V) and error (ε) terms in the probabilistic choice theory. The utility function can be denoted as Equation 2.1, while the deterministic portion (also known as observable portion) can be written as Equation 2.2. As shown in Equation 2.2, utility is a combination of driver characteristics (such as socio-demographic information, use of navigation systems, and attitudes toward dynamic message sign, DMS), alternative attributes (such as travel time and reliability of DMS information), and an interactive portion between them (such as route familiarity and past experience).

$$U_{rt} = V_{rt} + \varepsilon_{rt} \qquad \text{(Equation 2.1)}$$

$$V_{rt} = V(S_t) + V(X_r) + V(S_t, X_r) \qquad \text{(Equation 2.2)}$$

Where,
U_{rt}: Utility of route r to traveler t;
V_{rt}: Deterministic portion of utility of route r to traveler t;
ε_{rt}: Error associated with the utility of route r to traveler t;
$V(S_t)$: Portion of utility associated with the characteristics of traveler t;
$V(X_r)$: Portion of utility associated with the attributes of route r; and
$V(S_t, X_r)$: Portion of utility associated with the interactions between the characteristics of traveler t and the attributes of route r.

Given the deterministic part of utility, different structures can be defined to determine the probability of each alternative. These structures are basically derived from the assumptions regarding the distribution of the error part. Multinomial logit (MNL) is the most common form of discrete choice models being used in mode choice and route choice problems. A logit model assumes Gumbel distribution for the error term of the utility function. This approximation has calculation advantages to normal distribution, which lead to a multinomial probit model. MNL is formulated as Equation 2.3.

$$P_{rt} = \frac{e^{V_{rt}}}{\sum_{i=1}^{R} e^{V_{it}}} \qquad \text{(Equation 2.3)}$$

Where,
P_{rt}: Probability of route r being chosen by traveler t; and
V_{it}: Deterministic portion of utility of route i to traveler t, when there are R routes.

If there are only two choices to select, for example, decision between auto and transit as travel mode, binary logit (also known as logistic regression) is the suitable logit structure for binary choices. The probability function for route r between two choices for individual t can be written as Equation 2.4.

$$P_{rt} = \frac{1}{1 + e^{-V_{rt}}} \qquad \text{(Equation 2.4)}$$

While ordinary least squares (OLS) is the most common method to estimate unknown parameters in linear regression models (with continuous outcomes) to minimize error, maximum likelihood estimation (MLE) is the prevalent approach when investigating discrete choice outcomes. In simple words, MLE estimates the unknown parameters in a way that makes the observation most probable. The likelihood function for an MNL structure can be written as Equation 2.5. Because maximizing a sum function is simpler than such product function, we tend to maximize the logarithm of likelihood as shown in Equation 2.6, where $l(\theta)$ is the log likelihood function of θ. Therefore, θ represents the optimal probability configuration of individuals' route choice that maximizes their utility functions.

$$L(\theta) = \prod_{t=1}^{T} \prod_{r=1}^{R} P_{tr}^{Y_{tr}} = \prod_{t=1}^{T} P_{t1}^{Y_{t1}} P_{t2}^{Y_{t2}} \cdots P_{tR}^{Y_{tR}} \qquad \text{(Equation 2.5)}$$

$$l(\theta) = \sum_{t=1}^{T} \sum_{r=1}^{R} \log(P_{tr}^{Y_{tr}}) \qquad \text{(Equation 2.6)}$$

Where,
Y_{tr}: Model outcome for traveler t falls in route r; and
T: Total number of travelers.

In a categorical set of variables, a chi-squared test of independence can determine whether there is any association between any two factors drawn from a single population.

2.2   DEPARTURE TIME MODELS

It is known that how traffic congestion develops is a result of individuals' decision to travel; however, it must also be taken into account that how individuals decide *when* to make a trip can be influenced by real-time congestion information they obtain. This knowledge assists modelers in understanding how drivers adjust their departure time to avoid facing traffic congestion. This stage of travel demand models features the time of day distribution of trips. Travelers' departure time can be affected by policy variables, such as congestion pricing, parking regulations, high occupancy vehicle (HOV) restrictions, advanced traffic control strategies, and flexible work hours, other than the timely information they may receive through advanced traveler information systems (ATIS).

Discrete choice models, such as multinomial and nested logit, are the most common approaches in developing departure time choices to account for travel time uncertainty and activity duration. However, there is always compromise in discretizing an inherently continuous variable, such as time. The discrete choice approach treats the times near the boundaries as belonging to one interval or another; however, in reality, two close-spaced points should have similar attributes and utility. To overcome such shortcomings, a set of continuous-time models including hazard duration has been developed to consider the departure time choice spanning the entire day length.

There are also efforts of integrating departure choice models with activity duration or trip destination. Therefore, it is rational to assume that a traveler chooses a departure time to maximize the utility of both his or her trip and activity planned in trip origin or destination. These utility functions consist of trip and activity characteristics, such as trip time, trip cost, trip purpose (work trips versus non-work trips), travel mode, scheduling cost, clock time, activity duration, and demographics (gender, ethnicity, education level, age, and income).

Departure time choice for work trips is covered very well in the literature, while there are few studies about departure time choice for non-work trips. Small (1982) and many others modeled desired arrival time at work in the morning. Small found that travelers were willing to depart one to two minutes earlier or one-third to one minute later to save one minute of travel time. McCafferty and Hall (1982) modeled departure time choice to leave work in the evening using MNL. They concluded that neither travel time nor the socio-economic variables had a significant effect on departure time. Although most studies have utilized discrete choice models, especially, MNL, some recent studies used continuous methods. Mannering and Hamed (1990) used a joint discrete or continuous method. A discrete model was used for the decision of whether or not to delay departure, and then, the duration of the delay was modeled using a continuous Weibull survival function.

Example 2.1: An individual is thinking of three options for his everyday departure time to commute to work: 7:00 AM, 7:30 AM, or 8:00 AM. Historical traffic data provides relatively reliable travel time forecast for any departure time for this fixed origin–destination commute. The commute time forecasts for the aforementioned departure times are: 30, 45, and 40 minutes, respectively.

Assuming an MNL choice structure, determine the probability of the individual's utility for each option, whereby the utility function can be formulated as the following function. There is a disutility (or penalty) for any minute that the individual is late to work after 8:30 AM, and there is a utility for any minute that the individual can stay home later than 7:00 AM, creating time for personal tasks.

$$U_o = -0.1 - 0.03\,\mathrm{TT}_o - 0.05\,\mathrm{D}_o + 0.015\,\mathrm{MTS}_o$$

Where:
U_o: Utility of option O;
TT_o: Travel time in option O;
D_o: Arrival delay (after 8:30 AM) in option O; and
MTS_o: Morning time saving (after 7:00 AM) in option O.

Solution 2.1: Table 2.1 shows the delays and time savings for each option. The utilities can be calculated based on the presumed penalties (due to late arrival) and advantages (due to leaving home later).

$$U_1 = -0.1 - 0.03(30) - 0.05(0) + 0.015(0) = -1.0$$
$$U_2 = -0.1 - 0.03(45) - 0.05(0) + 0.015(30) = -1.0$$
$$U_3 = -0.1 - 0.03(40) - 0.05(10) + 0.015(60) = -0.9$$

Using the MNL function, the probabilities of all departure times can be calculated as follows. Option 3 earns the highest probability, and it can be concluded that the individual is more likely to depart at 8:00 AM than any other time. Although the utility values calculated can determine the most likely option, the probabilities would assist in developing aggregate predictions for a larger-scale problem.

$$P_1 = \frac{e^{U_1}}{\sum_{i=1}^{3} e^{U_i}} = \frac{e^{-1.0}}{e^{-1.0} + e^{-1.0} + e^{-0.9}} = \frac{0.368}{0.368 + 0.368 + 0.407} = 0.322$$

$$P_2 = \frac{e^{U_2}}{\sum_{i=1}^{3} e^{U_i}} = \frac{e^{-1.0}}{e^{-1.0} + e^{-1.0} + e^{-0.9}} = \frac{0.368}{0.368 + 0.368 + 0.407} = 0.322$$

$$P_3 = \frac{e^{U_3}}{\sum_{i=1}^{3} e^{U_i}} = \frac{e^{-0.9}}{e^{-1.0} + e^{-1.0} + e^{-0.9}} = \frac{0.407}{0.368 + 0.368 + 0.407} = 0.356$$

Table 2.1. Utility function attributes

Option	Departure time	Travel time (min.)	Arrival time	Arrival delay (min.)	Morning time saving (min.)
1	7:00	30	7:30	0	0
2	7:30	45	8:15	0	30
3	8:00	40	8:40	10	60

Example 2.2: If the individual discussed in Example 2.1 has the option of leaving home at any given time, what is the most probable departure time? Assume that the travel time linearly grows from the fast Fourier transform (FFT) (30 min.) at 7:00 AM to 45 min. at 7:30 AM, remains 45 min. until 7:50 AM, linearly degrades to the FFT at 8:20 AM, and maintains constant until evening peak hour.

Solution 2.2: The most probable departure time would be the time that maximizes an individual's utility. Travel time (TT), Delay (D), and Morning time saving (MTS) graphs are plotted in Figure 2.1 for a departure window of 6:00 AM to 9:00 AM. Given the TT for any departure time, arrival time can be estimated, and consequently, D would be determined for arrivals that occurred only after 8:30 AM. The MTS is applicable to departures after 7:00 AM only. Depending on both the departure and arrival times, the D function does not appear to be a simple linear chart. An individual's utility can be computed based on the equation provided in Example 2.1 and is plotted in Figure 2.2.

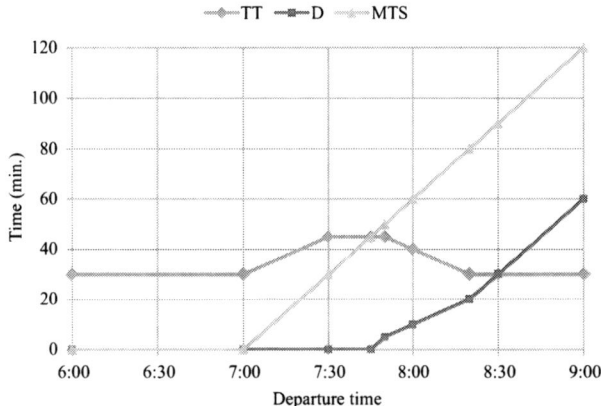

Figure 2.1. Travel time, delay, and saving time variations due to departure time.

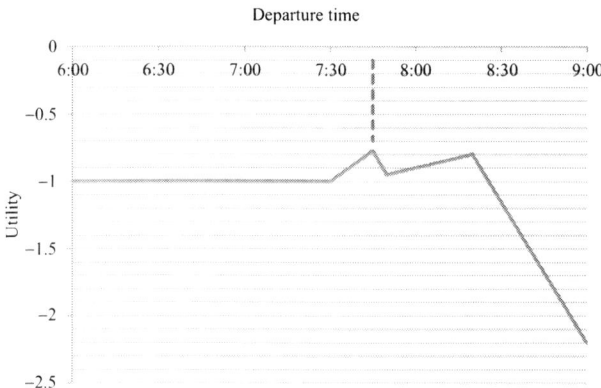

Figure 2.2. Utility as a function of departure time.

One way to interpret this is that with D having the highest weight in the utility function, if the individual can manage to arrive in his or her workplace by the threshold (8:30 AM), he or she would maximize his or her benefit of time saving while not incurring any delay penalty. In order to arrive by 8:30 AM, he or she must depart at 7:45 AM, which is the time that maximizes his or her utility. It is noteworthy that the utility graph in Figure 2.2 illustrates another maximum point at departure time 8:20 AM. This is due to the descending pattern of TT after 7:50 AM, although D outgrows the MTS when the trip starts after 7:45 AM.

2.3   DESTINATION CHOICE MODELS

Destination choice models are substitutes to gravity models (trip distribution), which are formulated as discrete choice models. Destination choice models provide a better behavioral basis for trip distribution than the gravity model by allowing for a wider range of explanatory variables than only distance or travel time. However, due to lack of data and unobserved factors, these models still struggle to explain the spatial distribution of travel. Destination choice modes can be used in tour- and activity-based models, as well as aggregate trip-based models.

Residential and work places are typically considered as exogenous in travel demand modeling. These locations are considered fixed in short run. The locations usually come from land use models that usually locate work places as given (primarily) and distribute residences around the work places.

The destination choice model formulation is as follows:

$$T_{ij} = P_i \frac{\exp(V_j)}{\sum_j \exp(V_j)}$$

where T_{ij} is number of trips between origin i and destination j, and V_j is the utility of choosing destination j. The utility contains the destination characteristics and attractiveness such as distance, number of retail stores, and restaurants , as well as the person's characteristics such as age, gender, and employment status.

2.4 MODE CHOICE MODELS

When the origin, destination, and time of travel are known, a traveler decides how to reach to the destination. Options can be stratified as follows:

- Non-motorized modes
  - Walk.
  - Bike.
- Motorized modes
  - Public: train, light rail, bus, and so on.
  - Private: auto (drive alone), carpool, and so on.
  - Semi-public: Taxi, ride share, and so on.

Mode choice models provide aggregate forecasts of travel modes across traffic analysis zones, or in other words, they estimate the proportion of travelers in any given origin–destination pair who would use each available mode of travel. Mode choice models essentially convert person-trip tables in conventional trip-based demand models into vehicle-trips. Mode choices are behavioral models; thus, they are sensitive to long-term changes in socio-economic characteristics of an area and to the potential upgrade or introduction of new transportation facilities, such as a new transit line, a bus-only lane, a new HOV lane, or an increase in toll rate. Mode choice model outputs can also be utilized to estimate parking demand for a park-and-ride service at a mass transit station. The main factors affecting one's decision as to which travel mode to choose can be classified into three major groups:

- Mode-specific attributes:
  - Travel time (walk time, wait time, in-vehicle time, total trip time, and so on)

○ Travel cost (transit fare, parking cost, fuel cost, auto depreciation cost, and so on)
○ Number of transfers (between transit lines or vehicles)
○ Comfort or convenience or service
• Traveler-specific attributes:
○ Demographics (age, gender, household size, employment status, value of time (VOT), income level, and so on)
○ Driving license status
○ Auto or bike ownership
○ Personal preferences and attitudes (e.g., attitude toward transit systems)
• Trip-specific attributes:
○ Travel distance (especially for walk and bike trips)
○ Time of day (safety concerns ins using public services, bike, or walk)
○ Trip purpose (work, non-work)
○ Trip chain or tour characteristics (shopping or family member pick-up followed by a work trip that increases the chance of using auto)
○ Orientation to the central business district (CBD)

Attributes can be either *generic*, such as traveler-related attributes, or *specific*, such as mode-related attributes. Generic attributes hold the same values across the modes; thus, they must be assigned different weights or parameters in the model. Specific attributes normally hold different values that distinguish them from one another (for instance, auto tends to be faster than bus); however, they can be calibrated with equal or unequal parameters. To clarify, while transit trip time can be higher than auto trip time in the model, the weight of transit time can also be higher than the auto time weight to imply that transit users might perceive the transit trip time as longer than what it actually takes.

The availability and accessibility of different travel modes vary among users. A drive alone option may not be available for travelers without a driver's license or those who do not own or lease a vehicle. Distances from trip origin and final destination to nearest transit stations also account for developing a feasible mode choice set for that trip. Public modes can be separated into walk-and-ride and drive-and-ride, depending on the accessibility of transit stops. However, the walk mode, either alone or with transit, always has a maximum walk length restriction.

Mode discrete choice models (multinomial logit or probit and nested logit) are the most common of mode choice analysis methods in a disaggregate level, when traveler-specific data is available. Mode choice is based on disaggregated behavioral models that reflect the actual choice process. The probability of choosing a mode of transportation is estimated using discrete choice models. When there are only two modes of transportation, the binomial logit model is used. Multinomial is the most popular approach and is used when there are several available modes of transportation. The nested logit model is used when classifying the modes into major modes, and then, each mode is classified further to other sub-modes. An example of a nested logit model is presented in Figure 2.3.

One main challenge with disaggregate mode choice models is data collection to perform model calibration and validation. Revealed preference (RP) survey data does not reveal information about unselected modes and are normally not adequately large enough to be split into calibration and validation sets. Stated preference (SP) surveys may not be an ideal representation of the decisions travelers actually make. Therefore, it is common in practice to transfer mode choice parameters from another geographic area with similar travel patterns.

In this nested structure, the selection of any mode in the lower level is contingent to the selection of its upper-level branch. For instance, the utility and the conditional choice probability of *bus* are calculated as follows:

$$U_{Bus} = U(bus \mid transit) + U(transit)$$

$$P_{Bus} = P_{bus \mid transit} \times P_{transit}$$

The nested logit formulation accounts for the interrelationship among choices in each nest. Let us assume μ_i represents the correlation among choices within the nest i. The probability of bus choice is formulated as follows, given the utility function of each mode, where *w.b* indicates the walk or bike mode (see Equation 2.7).

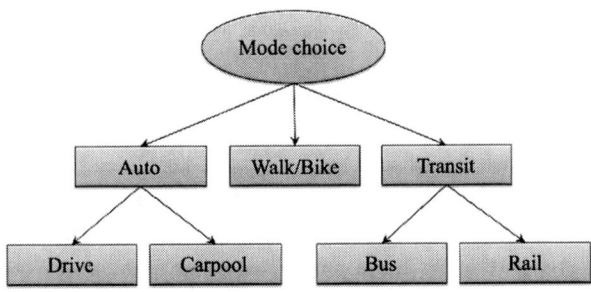

Figure 2.3. Transportation mode classifications for a nested logit model.

$$P_{Bus}$$

$$= \left(\frac{e^{\mu_{transit}U_{bus}}}{e^{\mu_{transit}U_{bus}} + e^{\mu_{transit}U_{rail}}} \right)$$

$$\times \left(\frac{\dfrac{1}{e^{\mu_{transit}}} Ln(e^{\mu_{transit}U_{bus}} + e^{\mu_{transit}U_{rail}})}{\dfrac{1}{e^{\mu_{transit}}} Ln(e^{\mu_{transit}U_{bus}} + e^{\mu_{transit}U_{rail}}) + \dfrac{1}{e^{\mu_{auto}}} Ln(e^{\mu_{auto}U_{drive}} + e^{\mu_{auto}U_{carpool}})} + \dfrac{1}{e^{\mu_{w,b}}} Ln(e^{\mu_{w,b}U_{w,b}}) \right)$$

(Equation 2.7)

In the United States, the dominant mode of travel is automobile (drive or car pool) due to the comfort associated with traveling by auto, which is not perfectly substitutable by transit comfort. The surveys show that modal choice behavior is marginally influenced by travel times and travel costs.

2.5   ROUTE CHOICE MODELS

Route choice models determine the route a traveler would choose, given the origin, destination, mode, and departure time of the trip. These models are traditionally based on random utility discrete choice frameworks. In this context, travelers choose a route from among several alternatives that maximize their utility. Unlike traffic assignment models, discussed in Chapter 4, route choice models are typically built in a disaggregate level. Advanced assignment models are based on the assumption of drivers' prior knowledge about the transportation network state. This assumption cannot be realistic because prior knowledge is a random variable and requires probabilistic models to describe the heterogeneity and learning process among users.

Generating realistic choice sets for individual travelers is considered a major challenge to developing fully behavioral route choice models. Incomplete knowledge of selected choice and unselected alternatives and an inability to generate a fully heterogeneous choice set may be the major causes. Behavioral route choice models can assist in better understanding of users' criteria for route preference other than simply travel time. New generations of commercialized information systems are envisioned to provide the best route, which is not necessarily the fastest, but perhaps, the most economic or scenic route, for instance. Drivers are observed to express route choice criteria other than travel time, such as minimizing stress or maximizing the aesthetic experience of travel.

Like many other choice problems in travel demand analysis—such as destination choice, departure time choice, and mode choice—route choice is frequently a pre-trip decision; however, for many reasons, travelers are required to make instant decisions along the way to cope with unexpected delays and beat traffic congestion. This type of common route decisions typically is formed under the high cognitive loads associated with multitasking (driving and decision-making) in real-world congested traffic environments, particularly in non-recurrent traffic conditions. In this case, the traveler's route decision is no longer a pre-trip plan. Route plan can be changed in an interactive environment that the driver has with the roadway, other drivers in the path, and the route guidance he or she receives.

New studies indicated that a realistic route choice model in the presence of information should include behaviors of both types of strategic and non-strategic thinkers as a function of past route experiences. Also, division of drivers into habitual and adaptive groups was demonstrated to be a key factor in calibrating a trip diversion model. Qian and Zhang (2013) have defined habitual travelers as those who are reluctant to switch to new routes and follow their pre-trip pattern, and adaptive travelers as those who are relatively responsive to travel information and inclined to use new routes occasionally.

2.6   CHOICE DATA COLLECTION

There have been two traditional methods of collecting choice data in transportation and many other behavioral sciences: SP and RP data. The SP data is usually gathered through survey questionnaires. The SP surveys are most widely used when the RP is unfeasible, such as investigating the popularity or potential demand of a new travel pattern (mode, route, and so on) that has not existed before, to gauge the future demand upon its supply.

The major criticism of SP versus RP is that travelers' attitudes elicited through SP do not necessarily coincide with their actual decisions or actions. The SP method has the inherent shortcoming of potential inconsistencies between the stated response to study questions and actual decisions in the real world, either due to the virtual nature of the SP context or the actual circumstance being somewhat different from what the SP scenarios present.

Two other major techniques are prevalent: field data and the driving simulator. Field experiments, a more precise but costly option, are not always feasible and safe to test various traffic information scenarios and environmental conditions. The driving simulator method necessitates a great deal of simulation efforts to develop a fairly realistic environment to test the question scenarios. The evolving driving simulator technology has been employed by transportation researchers for more than two decades to investigate drivers' controlled behavior for various driving conditions, including normal, fatigued, and drug-impaired.

Considering the cost, time, and practicality of each method, the best method is pertinent to the application. While SP fails to properly analyze drivers' route choice process under real-world complexity, the driving simulator technique supplies a more real environment to investigate the latent factors of route choice behavior. In an advanced driving simulator (DS), full information about all possible alternative routes is available. In addition, the use of various information provisions, controlled traffic, and environmental scenarios is not possible with other methods. The driving simulator is recognized to be a valid tool to explore the complex process of route decision-making because travelers' preferences change in the presence of information.

To investigate the relationship between drivers' behavior and DMS attributes, the majority of early studies were based on SP surveys because pursuing drivers in the real world is rather unsafe and impractical; however, some recent studies have applied the driving simulator technique to collect appropriate data for DMS effectiveness evaluation. Although SP outcomes do not guarantee that drivers' responses in the real world are identical to what they state, it is a valuable and less costly research tool for some specific research questions.

Factors such as time period, traffic conditions at off-ramps, and downstream traffic flow of the road are not distinguishable in typical SP-based driver behavior analyses. Therefore, the visible presence of queue, brake lights, and traffic congestion in the vicinity of the subject's vehicle, together, had significant effects on triggering the driver to divert. The actual diversion rate was found to be up to 80 percent less than what was stated in the SP questionnaire (Chatterjee et al. 2002).

BIBLIOGRAPHY

Avineri, E., and J.N. Prashker. 2003. "Sensitivity to Uncertainty: The Need for a Paradigm Shift." *Transportation Research Board Annual Conference.* Washington, D.C.

Ben-Akiva, M., and M. Bierlaire. 1999. "Discrete Choice Methods and their Applications to Short Term Travel Decisions." *Handbook of Transportation Science* 23, pp. 5–33.

Ben-Akiva, M.E., M.S. Ramming, and S. Bekhor. 2004. "Route Choice Models." In *Human Behaviour and Traffic Networks*, eds. M. Schreckenberg and R. Selten, 23–45. Berlin, Germany: Springer.

Bhat, C.R., and J.L. Steed. 2002. "A Continuous-Time Model of Departure Time Choice for Urban Shopping Trips." *Transportation Research Part B: Methodological* 36, no. 3, pp. 207–224.

Chatterjee, K., N.B. Hounsell, P.E. Firmin, and P.W. Bonsall. 2002. "Driver Response to Variable Message Sign Information in London." *Transportation Research Part C* 10, no. 2, pp. 149–169.

Ettema, D., and H. Timmermans. 2003. Modelling Departure Time Choice in the Context of Activity Scheduling Behavior. *Transportation Research Board Annual Conference.*

Hoffmann, J.P. 2004. *Generalized Linear Models: An Applied Approach.* Boston, MA: Pearson Education.

Koppelman, F.S., and C. Bhat. 2006. "A Self Instructing Course in Mode Choice Modeling: Multinomial and Nested Logit Models." Prepared for U.S. Department of Transportation, Federal Transit Administration.

Prato, C.G. 2009. "Route Choice Modeling: Past, Present and Future Research Directions." *Journal of Choice Modelling* 2, no. 1, pp. 65–100.

Qian, Z., and H. Michael Zhang. 2013. "A Hybrid Route Choice Model for Dynamic Traffic Assignment." *Networks and Spatial Economics* 13, no. 2, pp. 183–203.

Sheffi, Y. 1985. *Urban Transportation Networks: Equilibrium Analysis with Mathematical Programming Methods.* Upper Saddle River, NJ: Prentice-Hall.

Tian, H., S. Gao, D.L. Fisher, and B. Post. 2012. "A Mixed-Logit Latent-Class Model of Strategic Route Choice Behavior with Real-time Information." *Transportation Research Board Annual Conference.* Washington, D.C.

Xu, T.-D., L. Sun, and Z.-R. Peng. 2011. "Empirical Analysis and Modeling of Drivers' Response to Variable Message Signs in Shanghai, China." *Transportation Research Record, Journal of Transportation Research Board* 2243, pp. 99–107.

CHAPTER 3

Traffic Assignment Models

3.1 INTRODUCTION

Traffic assignment is the distribution of traffic demand onto transportation supply. In other words, traffic assignment is the assignment of origin–destination flows to transportation routes, based on factors that affect route choice. The purpose of traffic assignment is to find deficiencies in the existing transportation system in the current year as well as in the future, evaluate the effect of improvements to the existing system, and test alternative transportation system proposals.

There are two different equilibrium problems: the system optimal (SO) approach that minimizes the total system travel time over the planning horizon, and the user equilibrium (UE) problem that seeks time-dependent user path assignments that minimize travel time for each user. There are some other classifications of traffic assignment models such as:

- Stochastic/Deterministic: The route choice behavior of travelers is assumed to have a random component in the stochastic models.
- Path-based/Link-based: based on paths between origins and destinations or based on individual links that form paths.
- Flow-based/Vehicle-based: Flow-based models provide a snapshot of the network for each time interval (not a continuous record of network operation), while vehicle-based models require a starting point in time from which they proceed on a virtually continuous basis.

3.2 UE VERSUS SO

3.2.1 UE

The UE concept was defined by Wardrop's (1952) first principal: For each origin–destination (OD) pair, with UE, the travel time on all used paths is equal, and less than or equal to the travel time that would be experienced by a single vehicle on any unused path.

The UE problem is defined as time-dependent user path assignments that satisfy the extension of Wardropian UE conditions: No user can improve his or her travel time by unilaterally changing routes. Then, Beckmann et al. (1956) proposed a novel formulation for it. They proposed the first general mathematical formulation by applying Kuhn–Tucker condition. The formulation was a nonlinear programming problem based on behavioral assumptions of individual travelers to choose the minimum cost route over a congested network. However, no algorithm was proposed to solve their formulation until the late 1960s (Boyce 2007). The Beckmann formulation is as follows:

$$\min \sum_a \int_0^{V_a} S_a(x)\,dx \qquad \text{(Equation 3.1)}$$

subject to

$$V_a = \sum_i \sum_j \sum_r \delta_{ij}^{ar} X_{ij}^r$$

$$\sum_r X_{ij}^r = T_{ij}$$

$$X_{ij}^r \geq 0$$

where,

V_a: Number of vehicles (volume) per unit time on link a of the network;
$S_a(v_a)$: Travel time (cost) on link a;
X_{ij}^r: Number of vehicles from origin i to destination j on path r;
$\delta_{ij}^{ar} := 1$ if link a belongs to path r from origin i to destination j, 0 otherwise; and
T_{ij}: Given OD matrix

Travel time $S_a(v_a)$ is a function of link volume. It can be extended to be a generalized travel cost, depending on other variables as well as link volume.

The same year as Beckmann's formulation publication, Frank and Wolf (1956) proposed a convex combination algorithm for solving quadratic programming problems with linear constraints. LeBlanc et al. (1975) realized that the Frank–Wolf algorithm could be applied to solve the fixed demand traffic assignment problem. They applied the algorithm to a small network example and showed that the result is convergent. This algorithm has been widely used in determining equilibrium flows in static transportation network problems and has been adopted by different software developers. Peeta and Ziliaskopoulos (2001) and Friesz, Kwon, and Mookherjee (2007) proposed a dynamic version of this algorithm.

The Frank–Wolf algorithm is a solution to the following minimization problem with linear constraint:

$$\text{Minimize } Z(x) \qquad \text{(Equation 3.2)}$$

subject to

$$\sum_i a_{ij} x_i \geq b_j, \quad \forall j.$$

The algorithm is a feasible descent direction method for which at iteration $(n+1)$, a point x^{n+1} is generated from x^n so that $Z(x^{n+1}) < Z(x^n)$. The algorithmic step can be written as:

$$x^{n+1} = x^n + \alpha^n d^n$$

where d^n is a descent direction vector, and α^n is non-negative scalar known as the step size. The preceding equation means that at each point x^n, a direction d^n is identified along which the function is decreasing. Then, α^n determines the next point x^{n+1} will be along the direction d^n. To find a descent direction, the algorithm finds an auxiliary feasible solution y^n such that the direction from x^n to y^n provides the maximum drop with respect to a first-order approximation. Therefore, the descent direction $d^n = (y^n - x^n)$ is found using the following mathematical program that determines y^n.

$$\text{Minimize } \nabla Z(x^n)^T (y^n - x^n) = \sum_i \frac{\partial Z(x^n)}{\partial x_i}(y_i^n - x_i^n)$$
$$\text{subject to } \sum_i a_{ij} y_i^n \geq b_j, \quad \forall j \qquad \text{(Equation 3.3)}$$

where $\nabla Z(x^n)$ is the gradient of Z at x^n.

The next step is to find the step size α^n using the following line search problem.

$$\underset{0 \leq \alpha^n \leq 1}{\text{Minimize}} \quad Z[x^n + \alpha^n(y^n - x^n)]. \qquad \text{(Equation 3.4)}$$

After finding the descent direction and the step size, the next point can be generated using the following formula:

$$x^{n+1} = x^n + \alpha^n (y^n - x^n)$$

or   (Equation 3.5)

$$x^{n+1} = (1 - \alpha^n)x^n + \alpha^n y^n.$$

The last equation is a convex combination or weighted average of x^n and y^n. Starting with a feasible solution, the algorithm will converge after a finite number of iterations.

The UE formulation is based on several assumptions that might not always hold true. For example, it is assumed that travelers have full information about routes and travel times, they consistently choose the shortest path, and all travelers are identical in their behavior.

3.2.2 SO

Dafermos and Sparrow (1969) defined the concepts of UE versus SO. In the UE case, individual users decide based on their own interest, while in SO, they select routes according to what is optimal from a societal point of view. The SO problem is based on Wardrop's (1952) second principle: The average travel time is minimal.

The SO approach minimizes the total system travel time over the planning horizon. While UE formulations are descriptive in nature, the SO objectives are normative. The SO problem objective function minimizes the total system travel time (cost).

In UE, user travel costs on used paths for each OD pair are equalized, while in SO, marginals of the total travel cost on used paths for each OD pair are equalized.

The first mathematical programming approach to the SO problem was developed by Merchant and Nemhauser (1978). Their model considered the assignment of a known time varying OD trip pattern, and was formulated as a non-linear, non-convex, discrete time mathematical program. Most studies on SO traffic assignment can be accommodated in the following five classes of problems: (1) optimizing the departure patterns in a commuting corridor with a single route and one bottleneck, (2) minimizing total system cost for deterministic time-dependent OD flows from multiple origins to a single destination, (3) the joint assignments of peak-period users to departure times and routes in a corridor with a single destination,

(4) minimizing total system cost for stochastic time-dependent OD flows from multiple origins to a single destination, and (5) minimizing total system cost for deterministic time-dependent OD flows from multiple origins to multiple destinations. A typical SO formulation using mathematical programming is as follows (Peeta and Mahmassani 1995):

$$\text{Minimize} \sum_{\tau}\sum_{i}\sum_{j}\sum_{k}(r_{ijk}^{\tau}.T_{ijk}^{\tau})$$

or

$$\text{Minimize } [T(r_{ijk}^{\tau}, \forall i,j,k,\tau)]$$

subject to

$$r_{ij}^{\tau} = \sum_{k} r_{ijk}^{\tau}, \qquad \forall i,j,\tau,$$

$$\sum_{b} d^{tb} = \sum_{c} m^{tc} + I_n^t - o_n^t, \qquad \forall t,n,b \in B(n),c \in c(n),$$

$$x^{ta} = x^{t-1a} + d^{t-1a} - m^{t-1a}, \qquad \forall t,a,$$

$$x^{ta} = \sum_{k}\sum_{\tau}\sum_{i}\sum_{j}(r_{ijk}^{\tau}.\delta_{ijk}^{\tau ta}), \qquad \forall t,a,$$

$$T_{ijk}^{\tau} = \sum\sum[\delta_{ijk}^{\tau ta}\Delta], \qquad \forall i,j,k,\tau,$$

$$\delta_{ijk}^{\tau ta} = F[(r_{ijk}^{\tau}), \qquad \forall i,j,k\tau] \qquad \forall i,j,k,\tau,t,a,$$

$$d^{ta} = \sum_{k}\sum_{\tau}\sum_{i}\sum_{j} d_{ijk}^{\tau ta}, \qquad \forall t,a,$$

$$m^{ta} = \sum_{k}\sum_{\tau}\sum_{i}\sum_{j} m_{ijk}^{\tau ta}, \qquad \forall t,a,$$

$$I_n^t = \sum_{j} r_{nj}^t, \qquad \forall t,n \in I$$

$$o_n^t = \sum_{k}\sum_{\tau}\sum_{i}\sum_{c} m_{ink}^{\tau tc}, \qquad \forall t,n,\in J,c \in C(n),$$

$$\tau \le t,$$

$$\delta_{ijk}^{\tau ta} = 0 \text{ or } 1, \qquad \forall i,j,k,\tau,t,a,$$

all variables ≥ 0,

where,
i: Subscript for origin nodes;
j: Subscript for destination nodes;

n: Index for a node in the network;

a: Subscript for a link in the network;

k: Subscript for a path in the network;

τ: Superscript denoting departure time interval;

t: Superscript denoting current time interval;

T': Total duration for which assignments are to be made;

Δ: Length of a time interval (equal to T'/T);

r_{ij}^{τ}: Number of vehicles that depart from i to j in period τ;

r_{ijk}^{τ}: Number of vehicles that depart from i to j in period τ along path k;

T_{ijk}^{τ}: Experienced path travel time for vehicles going from i to j that are assigned to path k at time τ;

$\delta_{ijk}^{\tau t a}$: Time-dependent link-path incident indicator, equal to 1 if vehicles going from i to j assigned to path k at time τ are on link a in period t, that is,

$$[\delta_{ijk}^{\tau t a} = 1, \text{ if } r_{ijk}^{\tau} \text{ is on arc } a \text{ during period } t,$$
$$= 0, \text{ if arc } a \text{ does not belong to path } k,$$
$$= 0, \text{ if } \tau > t,$$
$$= 0, \text{ if } r_{ijk}^{\tau} \text{ is not on arc } a \text{ during period } t \ ],$$

$x_{ijk}^{\tau t a}$: Number of vehicles going from i to j along path k in period τ that are on link a at the beginning of period t; and

$d_{ijk}^{\tau t a}$: Number of vehicles going from i to j along path k in period τ that enter link a in period t.

The constraints represent the conservation of OD demand choices at the origin, conservation of vehicles at nodes, conservation of vehicles on links, the number of vehicles on a link in terms of path-vehicle assignments, definition of path travel times using the incidence variables, the definitional constraints for the number of vehicles entering and exiting links during various time intervals, the definitional constraints for the number of vehicles entering and exiting the network at a particular node and time interval, temporal correctness constraints that restrict the start (or departure) time constraint, restriction of the time-dependent incidence variables to take 0 or 1 values, and the non-negativity constraints.

As stated earlier, total travel times (cost) of used paths for each OD pair are equalized in UE, while marginals of the total travel times (cost) are equalized in SO. The UE and SO results turn out to be the same in low traffic flows. When the network becomes more congested and flow between origins and destinations increases, the UE and SO results diverge. This is because marginal travel time increases more rapidly than the travel time when flow is near capacity.

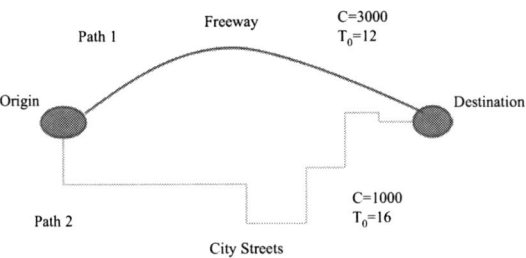

Figure 3.1. Example comparing UE and SO.

Example 3.1: Assume there are two paths, a freeway and city streets between an origin and a destination, as shown in Figure 3.1. The capacity of path 1 (freeway) and path 2 (city streets) are 3,000 and 1,000 vehicles per hour and the free flow travel times are 12 and 16 minutes, respectively. A flow of 8,000 vehicles per hour travels from the origin to the destination. The travel cost (time) is as follows.

$$S_a(V_a) = T_0 \left(1 + .15 \left(\frac{V_a}{C_a} \right)^4 \right)$$

The UE is reached when $S_1(V_1) = S_2(V_2)$. As shown in Figure 3.2, equilibrium occurs when path 1 has a volume of 6,154 and path 2, 1,846. The travel time on both paths is 43.87 minutes. Therefore, no user can be better off by switching from one route to the other. However, SO is reached when

$S_1(V_1) + \dfrac{dS_1(V_1)}{dV_1} = S_2(V_2) + \dfrac{dS_2(V_2)}{dV_2}$ and it happens when the volume of

path 1 is 6,001 and path 2 is 1,999. The travel time on path 1 is 40.82 and on path 2 is 54.32 minutes. Users on path 1 have considerably lower travel time than the ones on path 2, but the average travel time is minimum.

3.3   STATIC TRAFFIC ASSIGNMENT

Static assignment models assume that link flows and link travel times remain constant over the planning horizon. Thus, a matrix of steady-state OD trip rates are assigned to the network links. Although static equilibrium models are adequate for long-term planning analyses, they fail to capture the essential features of traffic congestion.

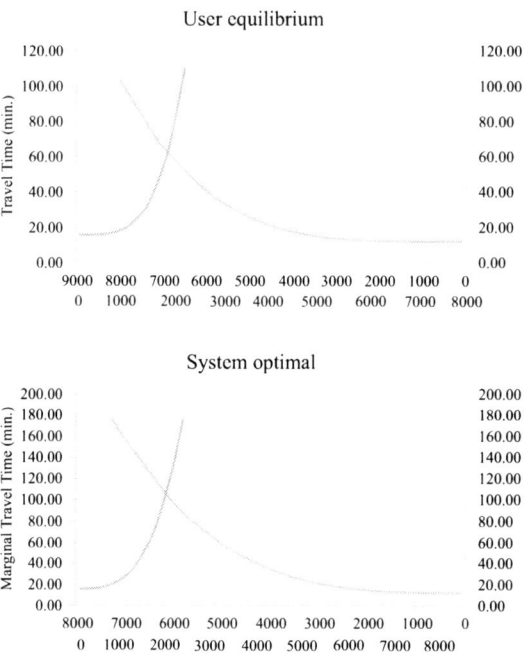

Figure 3.2. UE and SO solutions for Example 3.1.

In the static case, the link performance function (link travel time) is assumed to be positive, increasing, and convex. An unrealistic and restrictive assumption of these functions is that the travel time on each link is independent of flows on other links in the network. It is also assumed that link inflow and outflow are always equal, while traffic congestion and queuing occur when link inflow is greater than link outflow. The volume-to-capacity (V/C) ratio on a link may exceed 1. As it is impossible to have a flow greater than capacity, it can be interpreted that the assigned link volumes are demand rather than flow. When demand is greater than capacity, congestion will occur. The shortcoming of using V/C is that it does not directly correlate with any variable describing congestion (e.g., speed, density, or queue).

Static assignment models are inappropriate for real-time traffic control applications because congestion cannot be modeled adequately in static models. Although static traffic assignments (STAs) have been widely used for planning purposes and are adequate for capacity expansion projects, they are inadequate for evaluating policies for managing transportation systems and measuring environmental impacts related to system-wide travel demand. An improvement to the STA is the time-of-day STA that has been extensively used for travel demand modeling.

3.4   DYNAMIC TRAFFIC ASSIGNMENT

Dynamic traffic assignment (DTA) presents time-varying network variables and interactions using a behavioral approach. DTA can be used to evaluate many meaningful measures related to individual travel time (cost) as well as system-wide network measures for regional planning purposes.

Dynamic network assignment is a subject that has been considered by many researchers in the last three decades. The dynamic analysis of network flow patterns consists of three principal dimensions (Mahmassani 1990):

1. Time-dependent flow patterns within a given day: Almost all existing time-dependent traffic assignment models address single-user-class problems and can be classified into descriptive and normative models.
2. Day-to-day dynamics of peak period traffic flow: This is concerned with the evaluation of time-dependent flow patterns from day to day. This approach is appropriate for situations in which users adjust their departure times and route decisions from one day to the next and the system might not be in equilibrium. The stability of possible equilibrium under these conditions depends on the behavioral rules.
3. Real-time dynamics of flow patterns resulting from real-time decisions of motorists: This focuses on flow patterns resulting from the real-time decisions of travelers in response to supplied real-time information and perceived traffic conditions.

DTA could be an extension to Wardrop's UE by adding departure time.

Most existing time-dependent traffic assignment (TDTA) formulations assume convex, continuous, and non-decreasing link performance functions to represent link costs. This assumption simplifies the structure of the problem, but ignores essential aspects of the time-dependent nature of the problem. To minimize system-wide travel delays, it may often be useful to favor certain traffic streams or movements over others, which is probably not acceptable. Some mathematical programming SO/UE papers use the concept of a concave exit function specified as an upper bound on the number of vehicles exiting a particular link in a given period to model link congestion (e.g., see Li et al. [2000]). The behavior of traffic on a link is first-in-first-out (FIFO) and that creates a vexing difficulty in the solution of mathematical programming formulations. This problem does not arise in static assignment problems or in TDTA with a single destination. However, in TDTA with multiple destinations, vehicles on different paths that share one or more links may violate the FIFO protocol. For example, the downstream link along one path might be blocked, but not for another path. This problem occurs regardless of whether SO or UE models are being used. There are some techniques that have been proposed to address the FIFO problem in the literature (e.g., Peeta and Mahmassani [1995]).

Finding a dynamic user equilibrium (DUE) is challenging because each traveler's best route choice depends on the route choices of other travelers, who depart earlier, at the same time, or later. Transportation Research Circular (2011) states characteristics of a DTA solution as follows:

- Vehicles departing at different time slices are assigned to different routes.
- Vehicles departing at the same time slice for the same OD pair, but taking different routes should experience the same travel time.
- Experienced travel time is realized only at the end of the trip, not at the departure time.

DUE is found through an iterative process using an initial set of route choices and gradually improving them. An exact equilibrium may not be found in mid-size or large networks in a reasonable amount of calculation time. Therefore, an approximate equilibrium is reached in many current DTA models.

Figure 3.3 presents a general DTA algorithm.

As illustrated in Figure 3.3, the first step is network loading, which can be analytical or simulation-based. The analytical approach uses an exit function to predict traffic spreading in the network, while the simulation approach represents changes in traffic flow every few seconds. The next

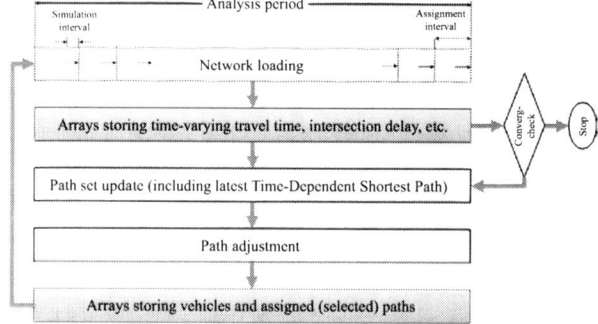

Figure 3.3. General DTA algorithm procedure (Transportation Research Circular (2011), p. 9).

step is path set update, which analyzes the results of the network loading. A time-dependent shortest path (TDSP) algorithm is utilized to find the shortest paths between each OD pair for each departure time period. In each iteration, the latest TDSP is combined with the ones calculated in the previous iterations for each OD pair and departure time period. The last step is path adjustment. The paths for which the flow should be adjusted (increased and decreased) and the amounts of adjustments are determined in this step. Some vehicles are selected to be rerouted. The algorithm iterates until convergence criteria is met. The older DTA algorithms use method of successive averages (MSA), which uses a predetermined fixed flow adjustment, while the newer algorithms apply relative gap which converges to the equilibrium faster than the MSA. The relative gap method stopping criteria is the difference between travel time of all used paths and the shortest path travel time divided by the shortest path travel time, to be less than a predetermined number.

We modified Example 3.1 to find DTA using the simple DTA in TransCAD. We determined 15 minutes intervals so the capacity changes to a quarter of hourly capacity, 750 and 250 vehicles per hour for paths 1 and 2, respectively. The flow of 8,000 vehicles per hour is subdivided into four time slots of 15 minutes, as presented in Table 3.1.

DTA gives flow for each path for every 15 minutes as shown in Table 3.2. The summation of the flow over one hour is very similar to the STA; however, it gives more detailed information of the variation of flow during the hour of study (every 15 minutes).

3.5   DETERMINISTIC VERSUS STOCHASTIC UE

The concept of stochastic user equilibrium (SUE) was developed and formalized by Daganzo and Sheffi in 1977. SUE is a generalization of the UE definition: Equilibrium is reached when no traveler believes that his or her travel time can be improved by unilaterally changing routes. Therefore, the assumptions of UE are relaxed by assuming travelers use their own

Table 3.1. Flow per 15 minutes for Example 3.1

	Flow (per 15 minutes)
Time slot 1 (7:00–7:15 AM)	1,700
Time slot 2 (7:15–7:30 AM)	1,900
Time slot 3 (7:30–7:45 AM)	2,300
Time slot 4 (7:45–8:00 AM)	2,100

Table 3.2. Comparison of different methods' results

Method	Path 1		Path 2	
	Flow	Time (min.)	Flow	Time (min.)
SUE	6,154	43.87	1,846	43.87
Static SO	6,001	40.82	1,999	45.32
Stochastic UE	6,141	43.61	1,859	44.64
Dynamic UE	6,158	43.96	1,842	43.63
Time slot 1	1,318	20.57	382	22.57
Time slot 2	1,464	25.08	436	27.08
Time slot 3	1,763	39.50	537	41.50
Time slot 4	1,813	31.26	487	31.26

perceived travel time rather than actual travel times. If perceived travel time is equal to actual travel time, then SUE results would be the same as the UE ones.

The deterministic user equilibrium problem is based on the assumption that travelers choose paths with minimum travel time (cost), which is fairly realistic. It also assumes that travelers have full and accurate information about travel times on all alternative paths, they consistently make correct decisions, and all travelers behave identically, which is not realistic. SUE relaxes some of these assumptions by introducing a random component to the model.

We ran SUE for Example 3.1 in TransCAD software. The results show that the volume on path 1 is 6,141 and path 2 is 1,859. Travel times are 43.61 and 44.64 minutes, respectively. The results are very close to the UE results, but travel times are not exactly the same, as it happens in the real world. Table 3.1 compares the results of different methods for Example 3.1.

3.6 MULTIMODAL TRAFFIC ASSIGNMENT

Four-step models are traditionally single modal (car), and other modes such as transit and freight are studied and assigned separately, without any interactions between different modes. Some researchers and practitioners have tried to capture and account for the effect of each mode on the other. However, the problem is not solved fully unless the assignment method changes. Multimodal traffic assignment is typically implemented in DTA rather than STA, which makes DTA models even more complicated.

The multimodal traffic assignment models consider the choices among different modes of transportation based on the costs of the modes. The link travel time functions employed in these models depend on the number of trips of modes, not the number of vehicles. Dafermos (1971, 1972) extended the Beckman's equilibrium solution to include multiclass models of traffic, in which link travel cost depends upon the entire link load pattern, rather than being solely dependent on the flow on that link only. Following is Dafermos' multimodal traffic assignment formulation. M is a set of users or vehicles of type (mode) m.

$$\min \sum_a \int_0^{V_a} S_a(x)\, dx$$

where,

$$S_a(x) = \sum_{m \in M} \sum_{a \in L} C_a^m(x_a^1, ..., x_a^k)$$

He solved the problem by reducing the optimization for multiclass user transportation networks to a single-class user.

REFERENCES

Beckmann, M.J., C.B. McGuire, and C.B. Winsten. 1956. *Studies in the Economics of Transportation.* New Haven, CT: Yale University Press.

Boyce, D. 2007. "Future Research on Urban Transportation Network Modeling." *Regional Science and Urban Economics* 37, pp. 472–481.

Boyce, D. 2013. "Beckmann's transportation network equilibrium model: Its history and relationship to the Kuhn–Tucker conditions." *Economics of Transportation* 2, no. 1, pp. 47–52.

Dafermos, S.C. 1971. "An extended traffic assignment model with applications to two-way traffic." *Transportation Science* 5, pp. 366–389.

Dafermos, S.C. 1972. "The traffic assignment problem for multiclass-user transportation networks." *Transportation Science* 6, pp. 73–87.

Dafermos, S.C., and F.T. Sparrow. 1969. "The Traffic Assignment Problem for a General Network." *Journal of Research of the National Bureau of Standards* 73B, pp. 91–118.

Frank, M., and P. Wolfe. 1956. "An Algorithm for Quadratic Programming." *Naval Research Logistics Quarterly* 3, nos. 1–2, pp. 95–110.

Friesz, T.L., C. Kwon, and R. Mookherjee. 2007. "A Computable Theory of Dynamic Congestion Pricing." *Transportation and Traffic Theory*, pp. 1–26.

Inoue, S., and T. Maruyama. 2012. "Computational Experience on Advanced Algorithms for User Equilibrium Traffic Assignment Problem and Its Convergence Error." *Procedia-Social and Behavioral Sciences* 43, 445–456.

Jeihani, M. 2004. "Enhancements to Transportation Analysis and Simulation Systems," Ph.D. Dissertation, Department of Civil and Environmental Engineering, Virginia Tech, Blacksburg, VA, https://vtechworks.lib.vt.edu/handle/10919/30092

Leblancl, J., E.K. Morlok, and W. Pierskall. 1975. "An Efficient Approach to Solving the Road Network Equilibrium Traffic Assignment Problem." *Transportation Research 9*, no.5, pp. 309–318.

Mahmassani, H.S., S. Peeta, T.Y. Hu, and A. Ziliaskopoulos. 1993. "Dynamic Traffic Assignment with Multiple User Classes for Real-Time ATIS/ATMS Applications." In *Large Urban Systems, Proceedings of the Advanced Traffic Management Conference*, 91–114. Washington, DC: Federal Highway Administration, U.S. Department of Transportation.

Meyer, M., and E. Miller. 2001. *Urban Transportation Planning: A Decision-Oriented Approach*. 2nd ed. New York, NY: McGraw-Hill.

Nemhauser, G.L., L.A. Wolsey, and M.L. Fisher. 1978. "An Analysis of Approximation for Maximizing Submodular Set Function-I." *Mathematical Programming*, no. 14, 265–294.

Peeta, S., and H.S. Mahmassani. 1995. "System Optimal and User Equilibrium Time-Dependent Traffic Assignment in Congested Networks." *Annals of Operations Research* 60, no. 1, pp. 81–113.

Peeta, S., and A.K. Ziliaskopoulos. 2001. "Foundations of Dynamic Traffic Assignment: The Past, the Present and the Future." *Networks and Spatial Economics* 1, no. 3, pp. 233–265.

Prato, C.G. 2009. "Route Choice Modeling: Past, Present and Future Research Directions." *Journal of Choice Modelling* 2, no. 1, pp. 65–100.

Sheffi, Y. 1985. *Urban Transportation Networks: Equilibrium Analysis with Mathematical Programming Methods*. Upper Saddle River, NJ: Prentice-Hall.

Transportation Research Circular 2011. *Dynamic Traffic Assignment A Primer*, E-C153.

Wardrop, J.C. 1952. "Some Theoretical Aspects of Road Traffic Research." *Proceedings, Institution of Civil Engineers Part* 2, no. 9, pp. 325–378.

CHAPTER 4

TRAVEL DEMAND MODELING APPROACHES

4.1 FOUR-STEP MODELS

Federal legislations in the 1960s required continuous, comprehensive, and cooperative (3C) urban transportation planning, which led to the four-step models.

The main characteristic of the four-step model is that it compartments the various aspects of travel demand. Because travel demand forecast is too complex to be modeled simultaneously, the four-step model breaks it into several issues and then combines it in a sequential order. The four-step model contains four sequential procedures: trip generation, trip distribution, mode choice, and trip assignment. The inputs of the model are traffic zones, travel surveys, socio-economic and demographic data, and transportation network and land use data. Using these inputs, the four-step model predicts the number of trips and loads travel volumes on the transportation network. This is done sequentially in four steps. First, using input data, the number of trips produced in and attracted to zones is determined (trip generation). Then, the generated trips are distributed among zones (trip distribution). The distributed trips then are split into different modes. Finally, the trips are allocated to feasible routes through a network. The overall process of the four-step model is presented in Figure 4.1.

Before performing a four-step model on a study area, the area needs to be delineated, zones need to be defined, and required input data in the household level and zone level needs to be acquired. The following section explains the data preparation process for the four-step model.

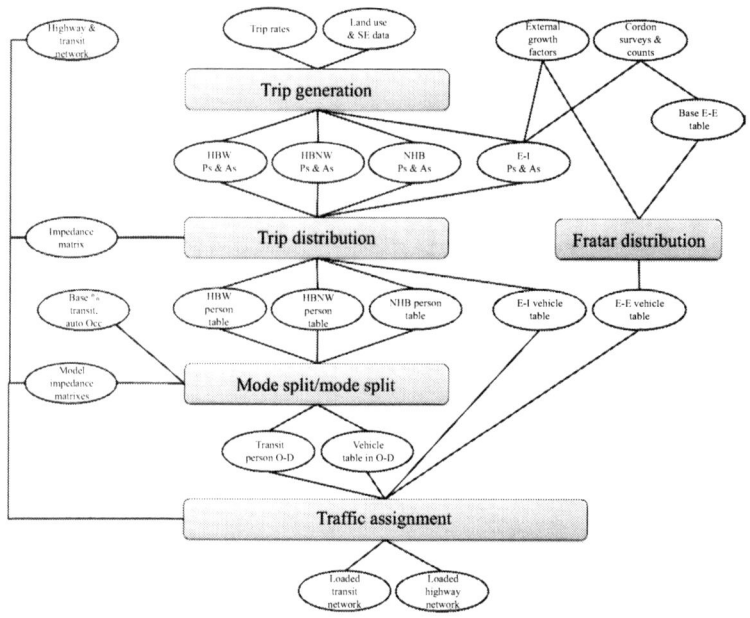

Figure 4.1. Four-step model process.

4.1.1 DATA PREPARATION

The first step is to specify the study area and its boundary. The study area must include all of the existing urban areas as well as the future developments in the next 20 to 30 years. It also needs to include the origin and destination of most of the trips in the area so that most of the trips will be internal ones and the data will be available. A good rule of thumb is to have at least 85 percent of the total study area vehicle miles traveled (VMT) accounted for by the internal–internal trips.

Trips in which both origin and destination are inside the study area are internal–internal trips. Trips with both origin and destination outside the study area are external–external trips. If the origin is inside the study area and the destination outside, it is an internal–external trip, and if the origin is outside the study area but the destination is inside, it is an external–internal trip.

The second step is to divide the study area into zones or traffic analysis zones (TAZs). Traditionally, TAZs are defined in such a way that a specific activity is in the majority. A TAZ can be residential, commercial, industrial, and so on. The boundaries of TAZs are usually manmade, such as roads, or nature-made, such as a river. TAZs, centroids, and centroid connectors were explained in Chapter 1. Figure 4.2 presents a small study

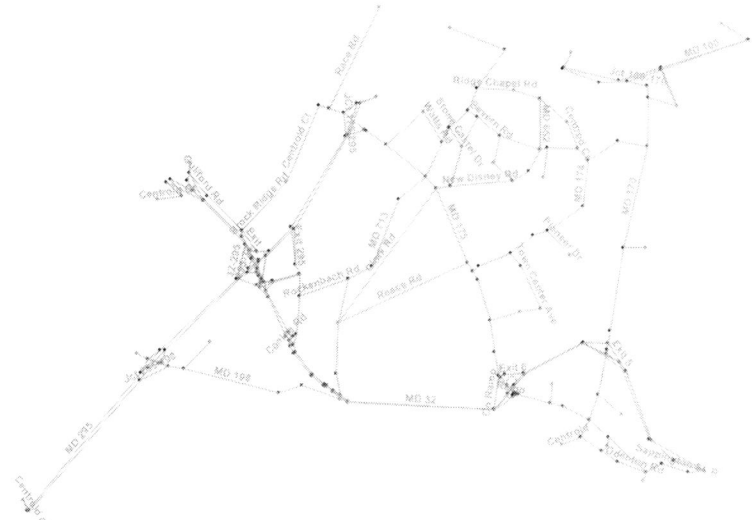

Figure 4.2. A small study area in Fort Meade, MD.

area with 28 TAZs and 263 links coded in a four-step model software, and Figure 4.3 presents a more detailed coded network for the same study area.

Transportation Research Board, TRB (2007) stated that the average number of TAZs for small metropolitan planning organizations (MPOs) (population less than 200,000) is 280, for medium MPOs (population of 200,000 to 1,000,000) is 870, and large MPOs (population over 1,000,000) is 1,760.

In order to develop a four-step model for a study area, the following data is needed for the base year as well as design years: transportation network, demographics, travel survey, land use, and economic growth.

The base year is the year for which the model is developed, calibrated, and validated. The base year has already happened and the data is available. The model tries to replicate the base year, and so, it needs to present similar results in traffic volumes and delays as the base year. Calibration and validation are performed to make the model as realistic as possible. Design years are future years that have not happened, for example, five years, 10 years, 15 years, 20 years, and 30 years from now. It is assumed that model parameters will stay unchanged during the years. Therefore, the calibrated or validated base model will be used to make the design year model by applying growth rates for population and economics, as well as applying the updated land use and network to replicate future changes in land use and roads.

Figure 4.3. Detailed road network of a small study area in Fort Meade, MD.

4.1.1.1 Transportation Network

A road network of the study area including all roads must be developed. Road characteristics such as length, number of lanes, capacity, direction, speed limit, and volume, need to be specified. Each zone is represented by its center of activity, called zone centroid. It is assumed that all vehicles originated from and are destined to the centroid of the zone. Zone centroid is connected to the roads via dummy links called centroid connectors. It is assumed that the centroid connectors have high capacity, so that there is no traffic congestion. In four-step models, local roads are not developed in the network because the origin and destinations are not individual houses, work locations, and other activity centers, they are only the centroid of each zone. Therefore, all local roads are substituted by the centroid connectors that connect the centroid to major roads. Centroids and centroid connectors are shown in red, TAZs in green, TAZ numbers in green, and road names in black in Figure 4.2.

Besides the road network, a transit network needs to be developed, and transit routes, stops, and schedule need to be determined in the network.

4.1.1.2 Demographics

Information about the population in the study area is obtained from census data. The data includes socioeconomic information about the people who live in the area. The population and demographic information is obtained for the base year. If the base year information is unavailable, the most recent information is adjusted to the base year.

The Census Bureau collects demographic data every 10 years. At the same time, an additional travel-related data (long form) is collected for one of six housing units in the United States. The Census Transportation Planning Package (CTPP) is developed based on the aforementioned data and provides socioeconomic, demographic, and travel information for transportation planners. (http://fhwa.dot.gov/planning/census_issues/ctpp/archives/index.cfm).

Public Use Microdata Samples (PUMSs) is a database collected by the Census Bureau (https://census.gov/programs-surveys/acs/data/pums.html), which provides information about each housing unit and each individual in the housing units. The data comes from the decennial census and the American Community Survey. PUMSs contain geographic units known as super-Public Use Microdata Areas (super-PUMAs) and Public Use Microdata Areas (PUMAs). Each PUMS file provides records for one in 1,000, 1 percent, and 5 percent of all housing units and each individual in those units in the United States.

Minimum population thresholds are set for PUMAs and super-PUMAs to maintain the confidentiality of the PUMS data. Super-PUMAs include a minimum population of 400,000 and consist of a PUMA or a group of PUMAs from the 5 percent state-level PUMS files. The 5 percent state-level files include PUMAs of populations of more than 100,000. The 5 percent files also will show corresponding super-PUMAs codes. Each state may include one or more super-PUMAs or PUMAs. PUMAs and super-PUMAs do not cross state boundaries. Large metropolitan areas may be subdivided into super-PUMAs and PUMAs.

The American Community Survey (ACS) is replacing the long form of the decennial census and will be the major source of census data for transportation planning. It samples a small percentage of the population every year. It sends about 250,000 survey questionnaires per month to U.S. households. The obtained data will be averaged over a three-year period.

4.1.1.3 Travel Surveys

Travel surveys are commonly conducted by MPOs. The questionnaire includes detailed information about socio-economic characteristics as well as travelers' attitude and diary. Their origin, destination, trip purpose, travel mode, time of travel, and so on for an entire day are asked in the questionnaire. The data is used to calculate trip production rate, trip attraction rate, average trip length, and trip length frequency, which are used in trip generation and trip distribution steps.

Three basic techniques used to conduct household travel surveys are the personal home interview, telephone interview, and mail-back survey. The personal home interview provides the most detailed data but is the most expensive and time-consuming method. Telephone interviews are less expensive and less time consuming and need fewer people to administer data collection. However, many people may refuse to participate. The mail-back survey has the lowest cost of all but also the lowest return rate.

4.1.1.4 Land Use

The required land use data for the four-step model comes from a land use model prepared by planning departments. Some departments utilize complex land use models, while many others use some form of a scenario-based approach or other judgmental techniques. A transportation modeler must obtain the most recent employment data for the county, urban area, or TAZ. If the data is not for the base year, the obtained data needs to be adjusted to the base year. If the data is not in TAZ level, conversion to TAZ level is needed. The next step is to identify special generators (e.g., military base, hospital, and university campus) and obtain their employment data. An example of a simple land use file includes population, number of households, median household income, number of employment by type (retail, office, industrial, and so on), school enrollment, and land acreage in the zone, for each TAZ.

Work place and special generator surveys, which are similar to household surveys, are used to obtain employment information. This data can be used in the trip generation step.

4.1.1.5 Population and Employment Forecast

Usually, transportation planners are provided a population and employment forecast (or control) value that is generated by state-level economic forecasts and then given to metropolitan and county planning agencies. In many states, the population and employment forecasts must conform to the specified total forecast, although the regional or local planning agency specifies the distribution of such forecasts within the study area.

Growth rate and the trend for population and economics are obtained from national and local sources and applied for future design years. Population for design years are forecasted in one of the following ways.

Constant Rate: A constant increase (decrease) per year or per five or 10 years. For example, if the population growth rate in Baltimore is 2 percent per year and the population for TAZ = 102 is 5,400 for the base year of 2015, then population for this TAZ will be 5,940 in year 2020 (2 percent increase per year, 10 percent increase in five years), 6,480 in 2025, 7,020 in year 2030, 7,560 in year 2035, and 8,640 in year 2045.

Ratio-Trend Method: Relate the population of a study area to the growth (positive or negative) ratio of the area's population and to the population of a larger area, for which an accepted population forecast exists. For example, the population of Baltimore City follows the population trend of the Baltimore–Washington area.

Cohort Survival Method: Add the effects of net natural increase and net migration to the existing population. For example, the population of Baltimore City was 621,000 in 2015. The 2016 population would be 621,000 plus the expected number of births minus the expected number of deaths plus the net expected number of migrations to or from the city.

Economic-Based Method: Relates population growth to a forecast of employment growth.

Compound Annual Rate: The forecasted population is calculated from the compound forecasted growth rate. For example, if the population growth rate in Baltimore is forecasted to be 2 percent per year fixed and the population for TAZ = 102 is 5,400 for the base year of 2015, then the population for this TAZ will be 5,962 in year 2020 $[(1 + 0.02)^{\wedge 5}]$; 6,582 in 2025 $[(1 + 0.02)^{\wedge 10}]$; 7,268 in year 2030 $[(1 + 0.02)^{\wedge 15}]$; 8,024 in year 2035 $[(1 + 0.02)^{\wedge 20}]$; and 9,781 in year 2045 $[(1 + 0.02)^{\wedge 30}]$.

Similarly, the growth rate for employment in each section is obtained and applied to find employment for each TAZ of the study area for all design years. However, employment forecasts are more difficult than population. Some forecasting methods are trend extrapolation, input/output analysis, and judgmental estimations.

Land use modeling will provide a land use file for each design year by incorporating the planned and predicted land use changes in the future. For example, if a new shopping center will be built in the next two years, it will affect the design years and be reflected in the land use file.

The road network and transit network also will be altered to reflect the future changes. For example, adding a lane to an arterial road in the next five years and adding a transit route next year should be reflected in each design year network.

4.1.2 TRIP GENERATION STEP

Trip generation is the first phase in the four-step model. The process of determining the number of trip productions and attractions associated with a given set of activities in a zone is called trip generation. The output of this step is unidimensional, P_i (number of trips produced in zone i) and A_j (number of trips attracted to zone j). Before finding trips attracted to and produced from each zone, trips are classified into different trip types. The most common trip types are home-based work (HBW), home-based other (HBO), and non-home-based (NHB). Depending on the study area and the objective of the study, some other trip types such as home-based school (HBS), home-based shop (HBSh), and so on can be used. Trips are assumed to be separate, and there is no trip chaining. For example, if a person goes from work to shopping and then goes home, the trip type will be NHB and HBO, rather than a chain of work-shop-home. The HBW trip is not considered in the trip generation process, and so, the HBW trip is underestimated. Figure 4.4 presents some trip types. Production and attraction are not necessarily the same as origin and destination because home is always a production source. For example, in Figure 4.4, when a person goes from home to work (a HBW trip), home is origin and also production and work is destination and also attraction. However, when the person goes from visiting a friend to home (a HBO trip), while home is a destination, it is again production, and while visit is an origin, it is an attraction. Therefore, in home-based trips, production and origin (attraction and destination) may be different. However, in NHB trips, origin and production (destination and attraction) are always the same.

There are two approaches to estimate the travel demand: aggregate level of zones and disaggregate level of households (sometimes person, employee, or establishment). In the aggregate level, the number of trips is

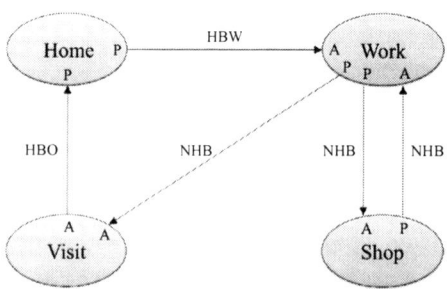

Figure 4.4. Trip types in four-step models.

assumed to be a function of zonal characteristics, while at the disaggregate level, it is assumed to be a function of households' characteristics. Because the four-step model is aggregate level, the trip generation results will be converted to aggregate zonal level. Two different methods can be used to generate trips: cross-classification techniques and linear regression. Cross classification is classification according to more than one attribute at the same time. Regression is a measure of the relationship between the mean value of one variable (output) and corresponding values of other variables (inputs). In all approaches, it is assumed that the relationship between factors (variables) is stable over time, and there is a linear relationship (or the relationship can be transformed to linear) between the dependent variable (trip rate) and independent variables. Independent variables are household characteristics such as household size, number of vehicles in the household, income, and number of licensed drivers in the household, or employment characteristics such as employment type, number of employees, and value of retail sales.

The number of trips produced is assumed to be dependent on car ownership, household income, household size, residential density, and distance from the central business district (CBD), and so on. The number of trips attracted to a zone is assumed to be a function of land use, employment, and other economic activities. After estimating the trip productions and trip attractions, a balancing is performed. The estimated number of trips produced at the household level should be equal to the number of trips attracted at the activity centers (Figure 4.5).

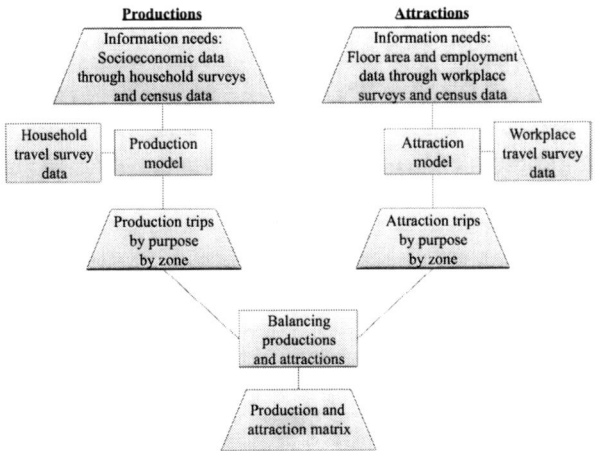

Figure 4.5. Balancing productions and attractions in trip generation.

4.1.2.1 Production Sub-model

Home is the major production source. People leave their homes to go to work, school, shop, and so on. Therefore, trip production is a function of the socio-economic characteristics of households, such as car ownership, family income, household size, occupation of household head, residential density, distance from the CBD, and number of licensed drivers in a household. Cross classification or a regression model can be used to model trip production.

In cross classification, households are grouped by household characteristics (such as number of vehicles in the household and family size) called strata, then the number of trips per household for each trip type is obtained from the data, and then, trip rates are calculated for each trip purpose. The trip generation rate for each specific cell is then the total number of trips in that cell divided by the number of households in that cell. Table 4.1 presents an example of trip rate using the cross-classification approach. A household of two people having one car would generate 5.7 trip productions in this example.

In the regression approach, trip rate (dependent variable) is estimated using linear regression as a function of household characteristics. Following is an example of a production model using regression.

$$P = 0.98 + 1.54 \ Veh + 0.023 \ Inc + 0.9 \ HH$$

where, P is the number of trips produced per household, Veh is the number of vehicles per household, HH is the household size, and Inc is the household income per $1,000. Therefore, a household of two people with one vehicle and an income of $50,000 will generate 5.47 trip productions using the preceding regression model. The calculated production rate using cross classification (5.7) or regression (5.47) is multiplied by the number of

Table 4.1. Number of trips produced per household per day (trip rate) for HBW trips

			Household size		
			1	2	3+
Number of vehicles		0	1	1.4	2.9
		1	2.6	5.7	6.8
		2+	4.2	6.6	9.0

households in each zone to find the trip production of the zone. A zone with 2,500 households will produce 13,675 trips using regression.

The data sources to find trip production rates are travel survey and census demographic data. Trip production (attraction) rates can be acquired from the ITE Trip Generation Handbook or other sources if the required data is not available for the study area. Those trip rates, however, are nationwide rates and may not be accurate for the study area.

4.1.2.2 Attraction Sub-Model

Trip attraction sub-models have received considerably less attention than have production sub-models. The techniques used to predict these trips are also less sophisticated. A lack of data is a major problem with trip attraction models. The same trip purposes used for the trip production models are used for the trip attraction models.

In cross classification, the most commonly used classification is employment type, which is divided into basic employment, retail employment, and service employment. In some cases, the classification is only retail and non-retail employment. The rates are typically derived from data aggregated over the entire zone, although in a number of cases, the area is divided into CBD, urban fringe, urban residential, suburban, and rural. Table 4.2 is an example of cross classification for attraction. The classification is for four area types of CBD, urban, suburban, and rural. For example, trip rate attracted to retail employment in a suburban area is 1.4 per employee and 9 for non-retail employment. Therefore, the number of trips attracted to an urban retail employment center of 200 employees is 280, and for a non-retail employment of 300 employees is 1,800.

In the regression approach, typically a simple regression model with only one or two variables is developed. Following is an example of a production model using regression.

Table 4.2. Number of trips per employment per day (trip rate) for HBW trips

		Employment type	
		Retail	**Non-retail**
Area type	CBD	1.3	2.2
	Urban	2.2	5.7
	Suburban	1.4	6.0
	Rural	1.4	2.4

$$A = -21.2 + 3.5 \text{ Emp}$$

where, A is the number of trips attracted per employment center and Emp is the number of employees.

An employment center of 500 employees would attract 1,728.8 (rounded to 1,729) trips per day.

Cross classification has some advantages over the regression model. Groupings are independent of the zone system of the study area. Unlike regression models, no prior assumptions of the shape of the relationship are required; therefore, it can easily accommodate non-linear relationships. However, large sample sizes are required; otherwise, cell values will vary in reliability. Also, it does not permit extrapolation beyond its calibration strata. Furthermore, there are no statistical goodness-of-fit measures for the model.

4.1.2.3 Special Generators

Some activity centers, such as schools and hospitals, whose trip generation characteristics are not fully captured by the standard trip generation module are considered special generators. In this case, specific information, such as the number of classes or number of beds, needs to be obtained to calculate trip rates.

4.1.2.4 Balancing

The final step in trip generation modeling is balancing productions and attractions. The estimated number of trips produced at the household level should be equal to the number of trips attracted at the activity centers, because each trip produced must be attracted somewhere. In the real world, the number of productions and attractions will not be exactly equal due to external trips.

Because there is a greater degree of confidence in the production models than in the attraction models, trip production totals are normally used as control totals, and attractions are scaled to productions. As a rule of thumb, the ratio of productions to attractions should not differ by more than 10 percent prior to balancing. Table 4.3 represents the balancing procedure for a region of five zones.

Because "Home" is always considered as production, no matter if it is an origin or a destination, balancing for all home-based trips is performed as aforementioned. Balancing cannot happen for each zone separately

Table 4.3. An example of balancing process in trip generation

Zone	Productions	Attractions
101	30	900
102	120	405
103	800	120
104	350	500
105	1,000	700
Total	**2,300**	**2,625**

Zone	Productions	Attractions * 0.87619
101	30	789
102	120	355
103	800	105
104	350	438
105	1,000	613
Total	**2,300**	**2,300**

because home is always production. For example, if a person goes from home (zone 101) to work (zone 103) and comes back from work (zone 103) to home (zone 101), there will be two productions for zone 101 and two attractions for zone 103. Thus, the summation of productions (attractions) for the two zones is two. However, as in NHB trips' origin is the same as production, and destination is the same as attraction, balancing must be performed for each zone separately. For example, the number of NHB production in zone 101 must be the same as the number of NHB attraction in zone 101.

After balancing, some error checking, validation, and calibration must be performed to make sure the results are fairly realistic and represent the study area.

4.1.3 TRIP DISTRIBUTION STEP

The second phase of four-step models is trip distribution, which specifies the number of trips between each pair of zones. The output of trip distribution is bi-dimensional T_{ij}, the number of trips between zone i and j. Trip distribution uses the outputs from trip generation and transportation system characteristics to distribute trips among zones. The traditional approach to trip distribution is a gravity model, which is adapted from Newton's gravitational law of physics in 1686 to explain the force between the planets and stars in the universe. The original formula is:

$$E_{ij} = G \frac{M_i M_j}{d_{ij}^2}$$
(Equation 4.1)

where,

E_{ij}: the gravitational force between bodies i and j
M_k: mass of body k ($k = i, j$)
d_{ij}: distance between bodies i and j
G: a constant

Transportation researchers adopted the preceding model, in which the number of trips between two TAZs is related to the attractiveness of the two TAZs and the distance between them. For example, for shopping, people tend to go to shopping centers that are closer to them and have a variety of options. That might not be completely true for work trips because a person already has his or her job. However, in choosing his or her home or job location, she or he considers the distance. A gravity model for trip distribution is as follows:

$$T_{ij} = P_i \frac{A_j / d_{ij}^b}{(A_1 / d_{i1}^b) + (A_2 / d_{i2}^b) + \dots + (A_j / d_{ij}^b) + \dots + (A_n / d_{in}^b)}$$
(Equation 4.2)

where,

P_i: number of trips produced by zone i
A_j: number of trips attracted to zone j
D_{ij}: distance (or travel time) between zone i and zone j
b: an empirically determined exponent that presents the average area-wide effect of distance between zones on trip interchange

A more general form of the gravity model is as follows:

$$T_{ij} = P_i \frac{A_j F_{ij} K_{ij}}{\sum_{j=1}^{n} A_j F_{ij} K_{ij}}$$
(Equation 4.3)

where, T_{ij} is the number of trips produced in zone i and attracted to zone j, P_i is the number of trips produced by zone i, A_j is the number of trips attracted to zone j, F_{ij} is the travel time factor, and K_{ij} is a specific zone-to-zone adjustment factor. The travel time factor can be any function of travel time, not necessarily $1/d_{ij}^b$. The preceding formula states that the percentage of trips produced by zone i allocated to destination zone j is dependent on both the attractiveness of zone j and travel time to that zone relative to the same features of all other attracting zones. Thus, increasing the attractiveness of a zone, such as building a new shopping center, increases its relative pull on the trip productions, and so, draws a greater proportion of these productions to itself. The matrix including T_{ij} values is called the trip interchange matrix or trip table.

The gravity model utilizes the travel time between TAZs. This travel time is calculated by finding the shortest path between zone centroids for each TAZ pairs. In the base year, the number of trips between each two zones is already known from expanding trip tables obtained from a travel survey to represent the whole population. Using the travel times between TAZs and trip generation output (P_i, and A_j), a transportation modeler determines travel time factors (F_{ij}) and zone-to-zone adjustment factors (K_{ij}) to reproduce trip tables (the number of trips between TAZs) of the base year. Assuming that F_{ij} and K_{ij} factors remain unchanged over the years, they are used in design years to find the number of trips between TAZs. Therefore, the trip distribution step for the base year is calibrating Equation 4.3 (estimating F_{ij} and K_{ij} values) using the observed data. However, in design years, the trip distribution step is estimating trip table (T_{ij} values) using the calibrated model.

Trip distribution based on a gravity model is easy to understand and applies in different study areas. It is responsive to changes in travel time between TAZs. It also accounts for the effect of trip purposes on zonal interchange.

4.1.3.1 Fratar Model

The Fratar model is the simplest trip distribution model, which was used before the gravity model introduction. It is a growth-factor model that simply extrapolates a base-year trip table based on a growth factor G to find trip distribution for design years. Because there is no information about external–external trips, the Fratar's model is used for these types of trips. The formula of the Fratar model is as follows.

$$T_{ij} = T_i \frac{t_{ij} G_j}{\sum_x t_{ix} G_x} \qquad \text{(Equation 4.6)}$$

where,

T_{ij}: The number of trips between TAZs i and j in the design year

T_i: Trips originating from TAZ i in the design year (t_i G)

t_i: Trips originating from TAZ i in the base year

G_x: Growth factor for TAZ x

t_{ij}: The number of trips between TAZs i and j in the base year

4.1.4 MODE CHOICE OR MODE SPLIT STEP

In this step, personal trip tables by trip purpose estimated in trip distribution are classified into different modes of travel such as auto, transit, and walk or bike. The input to this step is trip characteristics, travelers' characteristics, available modes, and travel costs by mode, as well as trip tables by trip purpose. The output is trip tables by mode, which is three-dimensional T_{ijm}, representing the number of trips between TAZ i and j using mode m.

Mode choice is used when the majority of people are choice riders, meaning people choose a mode among different modes based on the relative advantages of that mode over other modes. However, mode split is used when the majority is captive riders who have no choice and must use the only mode available to them. For example, there is no transit service in some areas; therefore, people have to use auto mode, or some people have no personal car or they are unable to drive, so they have to use transit mode. In this case, the characteristics and quality of the service offered do not matter. When using mode split, some transportation modelers apply mode split as a second step, right after trip generation, which is called trip-end model.

Example 5.1: Given the following utility functions for the three modes of car, bus, and walk or bike in a region, the probability and percentage of using each travel mode are calculated as follows. The cost for car is $2, bus is $1.25, and walk or bus is $0.25. Travel time of car is 10 min, bus 25 min, and walk or bike is 40 min.

$$U_{Car} = 2.2 - 0.2(cost_{Car}) - 0.03(travel\ time_{Car})$$

$$U_{Bus} = 1.3 - 0.2(cost_{Bus}) - 0.02(travel\ time_{Bus})$$

$$U_{W/B} = 0.5 - 0.2(cost_{W/B}) - 0.01(travel\ time_{W/B})$$

Example 5.2: Assume the mode choice problem in Example 5.1 is being analyzed in two stages. In stage one, the individual decides between motorized and non-motorized options, and if motorized is selected, a

Mode	Mode utility	Mode probability $P_{mn} = \dfrac{e^{U_{mn}}}{\sum_i e^{U_{in}}}$	Percentage
Car	$2.2 - 0.2 * 2 - 0.03 * 10 = 1.5$	0.617	62
Bus	$1.3 - 0.2 * 1.25 - 0.02 * 25 = 0.55$	0.239	24
Walk or Bike	$0.5 - 0.2 * 0.25 - 0.01 * 40 = 0.05$	0.145	14

second stage of mode choice process decides between car and bus. This is because the car and transit options might be correlated. Assume the correlation of 0.15 between the motorized modes, which is interpretable to $\mu_M = 1 - 0.15 = 0.85$.

According to Equation 3.7, nest choice probability and the probability of each mode conditional to the selection of its upper nest are calculated individually. New mode probabilities in the following table demonstrate slightly different results for each mode compared to the probabilities derived from an MNL structure. This deviation is a result of interdependency among alternatives. It is noteworthy that MNL can be viewed as an especial scenario of nested logit where $\mu = 1$. In this case, there is no need to cluster alternatives as their selections are fully independent of one another.

4.1.5 TRAFFIC ASSIGNMENT STEP

Traffic assignment is the fourth step that receives transportation network, trip table, and assignment rule as input and produces traffic flow or volume on each link of the network. Traffic assignment specifies four-dimensional t_{ijmr}, the number of trips between TAZ i and j using mode m taking route r. Figure 4.6 presents a schematic example of the four steps. Traffic assignment makes a relationship between the demand and supply side of traffic. Before applying traffic assignment, some procedures need to be performed as follows.

The 24-hour production (P) and attraction (A) trip tables by trip purpose and by mode need to be converted to origin (O) and destination (D) trip tables. As stated earlier, productions (attractions) are not necessarily the same as origins (destinations) because home is always considered as a

Nest	Mode	Modified utility	Conditional probability (Probability of mode choice, conditional on nest choice)	Nest LogSum	Nest probability	Mode probability
Motorized	Car	$1.5 * 0.85 = 1.275$	$e^{1.27}/(e^{1.27} + e^{0.47}) = 0.692$	$e^{(Ln(e1.27+e0.47)/0.85)} = 6.92$	$6.92/(6.92 + 1.05) = 0.868$	$0.69 * 0.87 = 0.600$
	Bus	$0.55 * 0.85 = 0.4675$	$e^{0.47}/(e^{1.27} + e^{0.47}) = 0.308$			$0.30 * 0.87 = 0.268$
Non-motorized	Walk or Bike	$0.05*$	1.00	$e^{0.05} = 1.05$	$1.05/(6.92 + 1.05) = 0.132$	$1 * 0.132 = 0.132$

* When there is only one alternative under a nest, modified utility and nest choice probability are the same as the ones calculated in MNL.

Trip generation   Trip distribution   Mode choice   Trip assignment

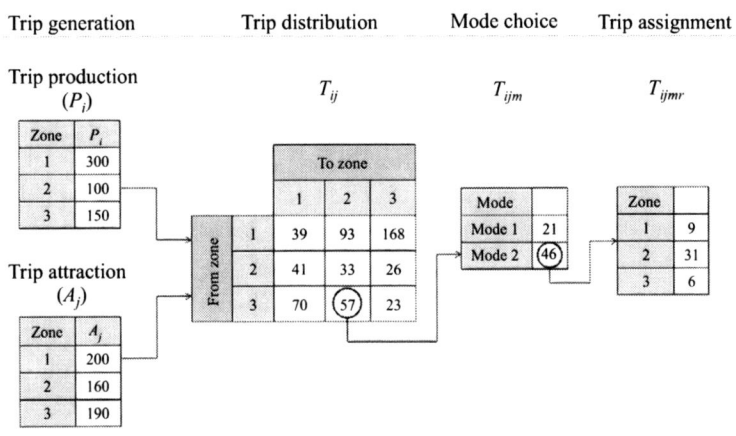

Figure 4.6. A schematic example of a four-step model.

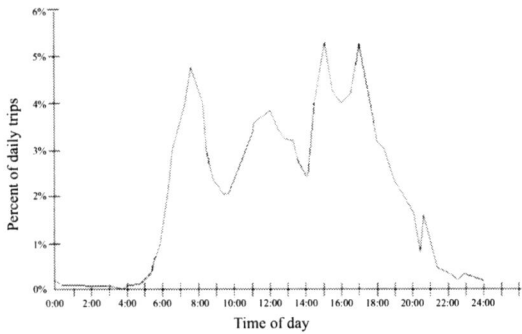

Figure 4.7. Daily trip distribution by time of day.

production source whether it is origin or destination. Therefore, a conversion is needed to change all trips to origins and destinations. It is assumed that half of the trips are from P to A directions and the other half are from A to P directions over the 24-hour period. Trip tables for all purposes are added together to find the total trips by mode. Then, the total trip table is added to its own transpose and divided by two. The result is an OD table.

The number of trips varies by the time of day and follows a pattern similar to Figure 4.7. The time of day usually is categorized to morning peak (one to three hours between 6:00 AM and 9:00 AM), afternoon peak (one to three hours between 3:30 PM and 7:00 PM), mid-day (between morning peak and afternoon peak, for example, 9:00 AM to 4:00 PM), and off-peak (after afternoon peak and before morning peak, for example, 7:00 PM to 6:00 AM). Usually, there are more trips in the afternoon peak than in the morning peak because travelers do more trip chaining on their trips from work to home (such as shopping and visiting), than in their morning trip from home to work. It is assumed that the time of day patterns remain the same in the design year as in the base year, which is not realistic.

The number of trips between each TAZ pair by each available mode is for 24 hours. Before modeling traffic assignment, some practitioners apply time of day modeling, while others apply it from the beginning (trip generation), and some others apply it after traffic assignment. Applying it after traffic assignment is the easiest way; the 24-hour link volume is multiplied by a time of day factor to estimate the link volumes. Time of day factors can be at least two factors of AM peak and PM peak. Then, a directional split is performed using observed traffic. It is assumed that peak factors and directional split are uniform for the study area, while some modelers use different factors for different road functional classification (freeway, arterial, collector, and local) or area type (CBD, urban, suburban, and rural). Time of day factors and directional split are calculated from 24-hour directional traffic counts for every 15 min.

If time of day modeling is applied before traffic assignment, factors must be developed from travel surveys (home interviews, external station surveys, truck travel surveys, and so on). The factors could be for all of the study area or they can be by area type and facility type. Before applying the factors to trip tables, the production–attraction (PA) trip tables need to be changed to origin–destination (OD) trip tables. Then, mode-specific trip tables are produced for each time of day by applying time of day factors, which could be two (only AM peak and PM peak) to all four trip tables (AM peak, PM peak, off-peak, and mid-day). Then, traffic assignment is applied for each time of day separately.

Time of day modeling is utilized to find traffic volumes and travel times by time of day, rather than average daily so that bottlenecks and overly congested areas are specified for congestion management programs.

After finding the OD trip table by mode of travel, it is time to develop the traffic assignment step. In this step, we assign vehicles to the road network using a static assignment technique such as all-or-nothing, iterative, incremental, user equilibrium, or stochastic user equilibrium assignment.

The rationale for trip assignment is to assign all trips to the route with minimum cost. The minimum cost routes are determined using shortest path algorithms, which will be explained later. The route having the least travel time is usually considered as the minimum cost route; however, some practitioners consider a combination of travel time and distance or a general cost containing monetary costs as well as time. The trip assignment procedure uses a link performance function to calculate the travel time on each link.

4.1.5.1 Link Performance Function

The link performance function presents the relationship between link travel time and link volume. It originates from the familiar speed–flow relationship in traffic engineering. As shown in Figure 4.8, when flow increases, speed tends to decrease after an initial period of little change; when flow approaches capacity, the rate of reduction in speed increases. Maximum flow occurs at capacity, and by forcing volumes beyond this value, a stable region with low flows and low speed is reached. For practical reasons, this type of relationship is considered in terms of travel time per unit distance versus flow in traffic assignment. This travel time–flow relationship is called link performance function and is presented in Figure 4.9.

There are various link performance functions in the literature such as the Bureau of Public Roads (BPR) and modified BPR link performance function, and Davidson's, CATS, and Toronto link performance functions.

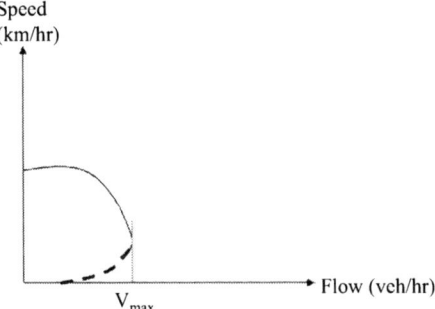

Figure 4.8. Typical speed–flow relationship.

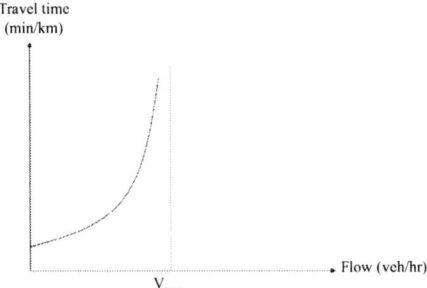

Figure 4.9. Travel time–flow relationship.

4.1.5.1.1 BPR Link Performance Function

BPR is the most common link performance function developed by the U.S. Bureau of Public Roads in 1964 as follows.

$$t_l(V_l) = t_{0l}\left(1 + \alpha\left(\frac{V_l}{C_l}\right)^{\beta}\right) \qquad \text{(Equation 4.9)}$$

where $t_l(V_l)$ is the average travel time on link l with volume V vehicles on it, t_{0l} is the free-flow travel time on link l, C_l is the capacity of link l, and α and β are model parameters. Typically, $\alpha = 0.15$ and $\beta = 4$ (the original values found by BPR), which assumes a level of service (LOS) of C. When volume (flow) on the link is small, link travel time is almost the same as free-flow travel time. Travel time increases slowly by volume until it reaches close to capacity. Travel time increases dramatically when volumes reaches capacity (Figure 4.10).

Some modelers use different values for α and β for each road type. The values for α and β are acquired by calibration and validation of Equation 4.9 using the observed data of the study area.

Some modelers modified BPR by defining two different functions for V/C less than 1 and greater than one.

$$t_l(V_l) = t_{0l}\left(1 + 0.15\left(\frac{V_l}{C_l}\right)^{4}\right) \quad V/C \leq 1 \qquad \text{(Equation 4.10)}$$

$$t_l(V_l) = t_{0l}\left(1 + 0.1\left(\frac{V_l}{C_l}\right)^{4}\right) \quad V/C > 1$$

This can be further modified by using different parameter values for each road type.

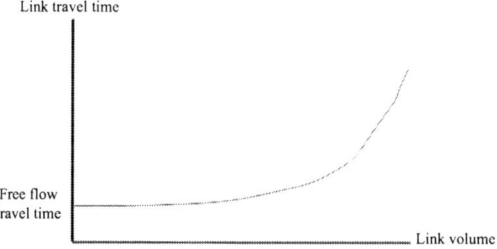

Figure 4.10. BPR link performance function.

Although BPR is very simple and has been widely used, it has some drawbacks. If volume is far lower than capacity, BPR (especially for high β values) gives free-flow travel time for the link, regardless of its traffic volume. BPR function works well for $V/C \leq 1$, but when $V/C > 1$ and β is high, the calculated travel time converges to infinity. This may cause slowing down convergence by giving excessive weight to overloaded links with high β values, and it can cause numerical problems such as overflow conditions. Figure 4.11 presents the BPR link performance function with different values of β. Part A shows the cases of $V/C \leq 1$, while part B shows the cases of $V/C > 1$.

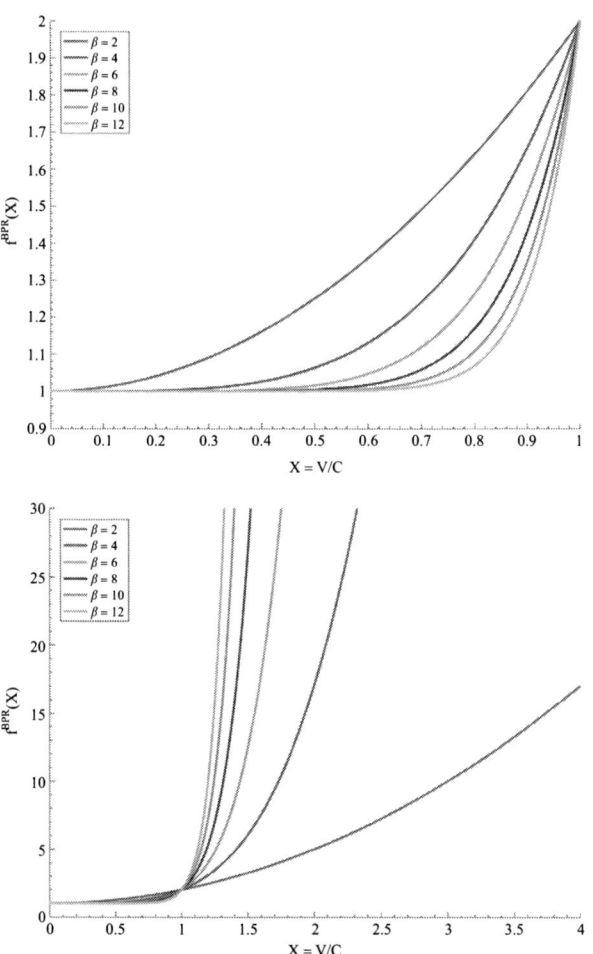

Figure 4.11. BPR link performance function with different β values.

4.1.5.1.2 Conical Link Performance Function

Conical functions were developed by Spiess (1990) to address some disadvantages of BPR. The results of conical and BPR are very similar when $V/C \leq 1$, as shown in Figure 4.12. However, they are different from BPR in the case of $V/C > 1$. The formula is as follows:

$$f^c(x) = 2 + \sqrt{\beta^2(1 - V/C)^2 + \gamma^2} - \beta(1 - V/C) - \gamma$$
$$\lambda = \frac{2\beta - 1}{2\beta - 2}$$

(Equation 4.11)

β: any number larger than 1.

4.1.5.1.3 Davidson's Link Performance Function

This function is proposed by Davidson (1966) based on the queuing theory as follows. Unlike BPR, Davidson's is asymptotic to capacity flow.

$$t_l = t_{0l}\left[1 + J\left(\frac{V_l}{C_l - V_l}\right)\right]$$

(Equation 4.11)

Parameter J determines the shape of the curve as shown in Figure 4.13.

4.1.5.1.4 Akcelik Link Performance Function

There are some difficulties with calibration of Davidson's function. Therefore, several scholars proposed modifications to the Davidson's function to obtain finite values of travel time for flows near and above capacity. Akcelik (1991) proposed a new link performance function as an alternative to Davidson's to address calibration problems. The function is proposed in both steady-state and time-dependent. The steady-state one has the same formula as Davidson's; however, parameter J is different. Equation 4.14 presents parameter J.

$$J = K'/(C\, t_0)$$

(Equation 4.12)

where k' is delay parameter.

Figure 4.14 compares Akcelik and Davidson's functions.

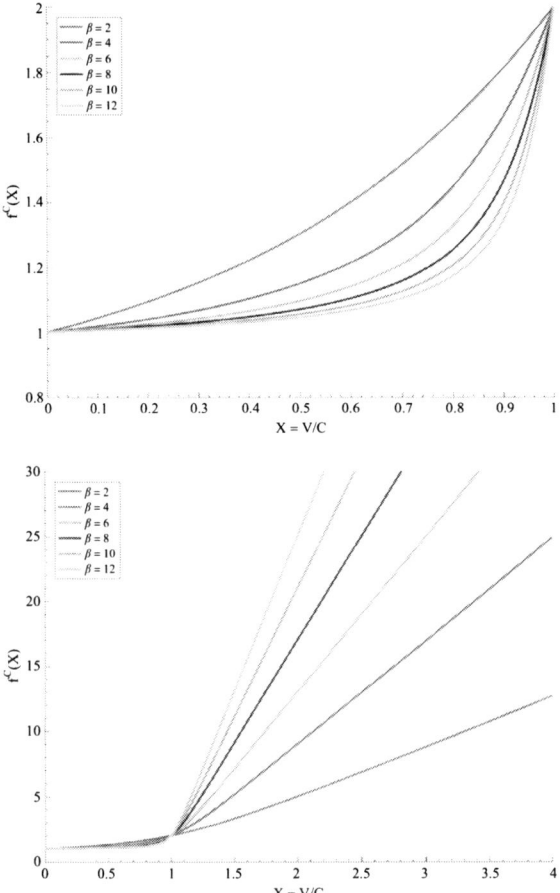

Figure 4.12. Conical link performance function with different β values.

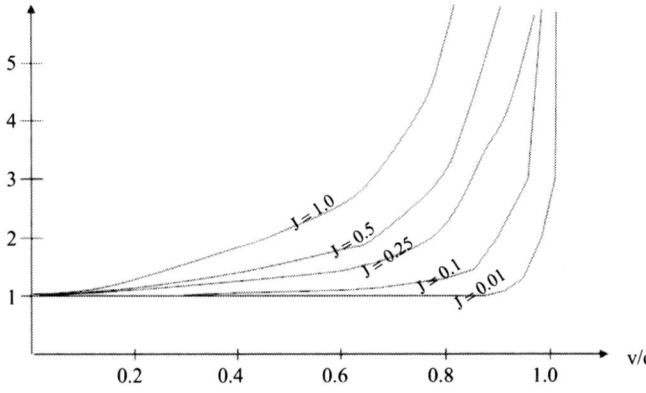

Figure 4.13. Davidson's link performance function (Yousef Sheffi 1985).

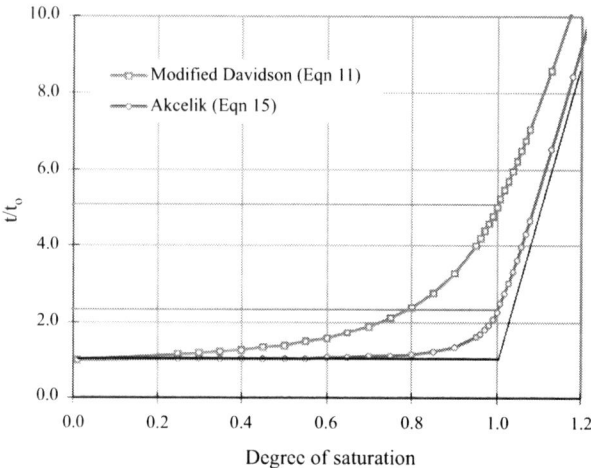

Figure 4.14. Akcelik versus Davidson's link performance function (Akcelik 2000).

4.1.5.1.5 Toronto Link Performance Function

This function assumes that travel time has a direct relationship with link characteristics such as speed limit, number of signalized intersections, and area type, in addition to link volume. Figure 4.15 presents an example of this function.

4.1.5.2 All-or-Nothing Assignment

All-or-nothing assignment is the easiest assignment technique, in which the shortest path between each origin and destination is specified, and then, all volume is assigned to the shortest path. However, when assigning, the shortest path would not be the shortest path due to excessive volume, and alternative paths would be shorter. Oscillating from one path to the other will not solve the problem. Therefore, other methods are provided and utilized in the literature. We follow the assignment methods using a simple example. Assume there is an OD with three possible paths: A (travel time 10 min, capacity 1,000 vehicles per hour), B (travel time 15 min., capacity 2,000 vehicles per hour), and C (travel time 17 min., capacity 1,500 vehicles per hour), as presented in Figure 4.16. The number of vehicles to be assigned from O to D is 5,000. As the shortest path is A, all 5,000 vehicles are assigned to it. Using the BPR link performance function, the travel time of path A is 947.5 min. as follows, while travel time on paths B and C are 15 and 17 min., respectively.

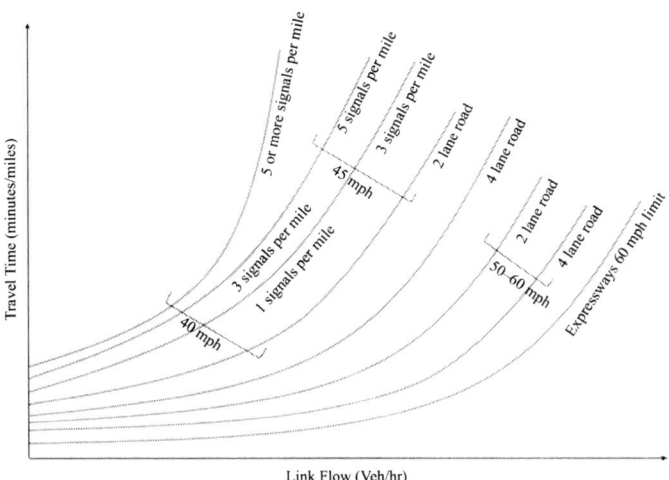

Figure 4.15. Toronto link performance function.

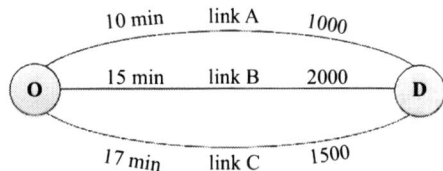

Figure 4.16. Traffic assignment example.

$$t_l(V_l) = t_{0l}\left(1 + 0.15\left(\frac{V_l}{C_l}\right)^4\right) = 10(1 + 0.15(5000/1000)^4)$$

$$= 947.5 \text{ minutes}$$

Path A is extremely congested. If we assign all 5,000 vehicles to the next shortest path (B), the same problem would happen. All-or-nothing assignment is single-path assignment, while other assignments are multi-path assignment techniques, in which flow is assigned to different paths. Some multi-path assignments procedures are as follows.

4.1.5.3 Iterative Assignment

It assigns the traffic into paths in four iterations by computing the weighted travel time consisting of 0.75 of the previous travel time and 0.25 of the current travel time and performing an all-or-nothing assignment in each step. Table 4.4 demonstrates iterative assignment for the previous example.

4.1.5.4 Incremental Assignment

In incremental assignment technique, the flow of each OD pair is divided into n (usually four) equal parts. An all-or-nothing assignment is performed for each part and the link travel time is updated after each assignment. The link volumes are summed after assigning the nth section. Incremental assignment for the previous example is as follows (Table 4.5).

4.1.5.5 User Equilibrium Assignment

User equilibrium is reached when no traveler can improve his or her travel time by switching paths. User equilibrium is the most common method utilized in four-step models. Please see Section 4.2 for more information. Following is the stepwise algorithm to solve the user equilibrium problem using Beckmann's formulation and the Frank–Wolf algorithm.

1. Compute the travel time on each link, $S_a(v_a)$, that corresponds to the flow v_a in the current solution.
2. Find the shortest path from each origin to all destinations by using the travel times from step 1.
3. Assign all trips from each origin to each destination to the shortest path (all-or-nothing assignment), w_a.
4. Combine the current solution (v_a) and the new assignment (w_a) to obtain a new current solution $(v'_a = (1 - \lambda)v_a + \lambda W_a)$ by using a value λ selected through a one-dimensional search to minimize the objective function

$$\sum_a \int_0^{v'_a} S_a(x)\,dx$$

5. If the solution has converged sufficiently, stop; otherwise, return to step 1.

Table 4.6 presents the user equilibrium assignment procedure for the previous example.

Comparing the three multi-path assignment methods (Table 4.7) shows that user equilibrium has minimum objective function, and travel

Table 4.4. Iterative assignment example

Iteration	Step	Link A		Link B		Link C		Equilibrium objective function
		Flow	Time	Flow	Time	Flow	Time	
Initial solution	–	5,000	947.5	0	15.00	0	17.00	
1	2	0	–	5,000	–	0	–	
	4	0	713.13	5,000	36.97	0	17.00	
2	2	0	–	0	–	5,000	–	
	4	0	537.34	0	31.48	5,000	95.70	
3	2	0	–	5,000	–	0	–	
	4	0	405.51	5,000	49.33	0	76.03	
Mean flows and corresponding travel time		1,250	13.66	2,500	20.49	1,250	18.23	

Table 4.5. Incremental assignment example

Iteration	Step	Link A		Link B		Link C		Equilibrium objective function
		Flow	Time	Flow	Time	Flow	Time	
0		0	10.00	0	15.00	0	17.00	
1	Assignment	1,250	–	0	–	0	–	
	Sum, Time	1,250	13.66	0	15.00	0	17.00	
2	Assignment	1,250	–	0	–	0	–	
	Sum, Time	2,500	68.59	0	15.00	0	17.00	
3	Assignment	0	–	1,250	–	0	–	
	Sum, Time	2,500	68.59	1,250	15.34	0	17.00	
4	Assignment	0	–	1,250	–	0	–	
	Sum, Time	2,500	68.59	2,500	20.49	0	17.00	

Mean flows and corresponding travel time

times on alternative paths are the same. Therefore, no user can improve his or her travel time by switching to another path.

4.1.5.6 Shortest Path Algorithm

All the aforementioned assignment techniques allocate vehicles to the shortest paths. A common shortest path algorithm is Moore's algorithm. This algorithm is fast because it reduces the computational time by developing a minimum tree from the origin to all other points in the network. It starts from the origin and connects it to all adjacent nodes and calculates the time to get from origin to each of the nodes. Then, it starts over from the closest node to the origin (minimum time). If a node is reached by a shorter time, the path with the longer time is removed from the tree. This procedure continues until all nodes are visited. The following example shows how the algorithm works. Assume that there is a small network of nine nodes. The travel time of each link is given in Figure 4.17a. The first step is to find nodes adjacent to the origin node 1 and find the time it takes to get to those nodes (Figure 4.17b). Node 4 is the closest node to the origin (node 1). Adjacent nodes to node 4 are nodes 5 and 7 (backtracking is not permitted). The corresponding travel times are travel time from node 1 to node 4 plus the travel times from node 4 to node 5 (two min) and to node 7 (three min), respectively (Figure 4.17c). The node closest to the origin is node 2 now. Proceeding from node 2 to node 3, travel time is the time from origin to node 2 plus travel time from node 2 to node 3, a total of five min (d). Likewise, the time to node 5 is five min. Travel time from origin to node 5 via path 1–4–5 is less than that via path 1–2–5. Therefore, the link 2–5 is deleted from the tree. This procedure continues until all nodes are visited and the final tree is built as presented in Figure 4.17.

The shortest path from node 1 to node 5 is 1–4–5 with total travel time of two min. Similarly, the shortest path from node 1 to node 9 is 1–4–5–8–9 with total travel time of seven min.

4.1.6 FEEDBACK

In the traditional four-step models, each step is treated serially and independently of the other steps, and the output of each step is passed to the next level. It is assumed that a hierarchy of travel choices exists: travelers first decide whether or not to travel, then which destination to choose,

Table 4.6. User equilibrium assignment example

Iteration	Step	Link A Flow	Link A Time	Link B Flow	Link B Time	Link C Flow	Link C Time	λ	Equilibrium objective function
1		5,000	947.5	0	15.00	0	17.00	−	
2	2	0	−	5000	−	0	−	.625	80,959
	4	1,875	28.50	3125	28.50	0	17.00		
3	2	0	−	0	−	5000	−	.244	74,412
	4	1,417	16.06	2363	19.38	1220	18.12		
4	2	5,000	−	0	−	0	−	.035	74,223
	4	1,543	18.50	2280	18.80	1177	17.97		
5	2	0	−	0	−	5000	−	.02	74,196
	4	1,512	17.84	2234	18.50	1254	18.24		
6	2	5,000	−	0	−	0	−	.007	74,189
	4	1,536	18.36	2219	18.41	1245	18.21		
7	2	0	−	0	−	5000	−	.005	74,188
	4	1,529	18.19	2207	18.34	1264	18.28		
8	2	5,000	−	0	−	0	−	.0001	74,187
	4	1,532	18.19	2205	18.33	1263	18.28		

Table 4.7. Comparison of different assignment methods

Iteration	Link A		Link B		Link C		Equilibrium objective function
	Flow	Time	Flow	Time	Flow	Time	
Iterative assignment	1250	13.66	2500	20.49	1250	18.23	75,220
Incremental assignment	2500	68.59	2500	20.49	0	17.00	94,543
User equilibrium assignment	1532	18.27	2205	18.33	1263	18.28	74,187

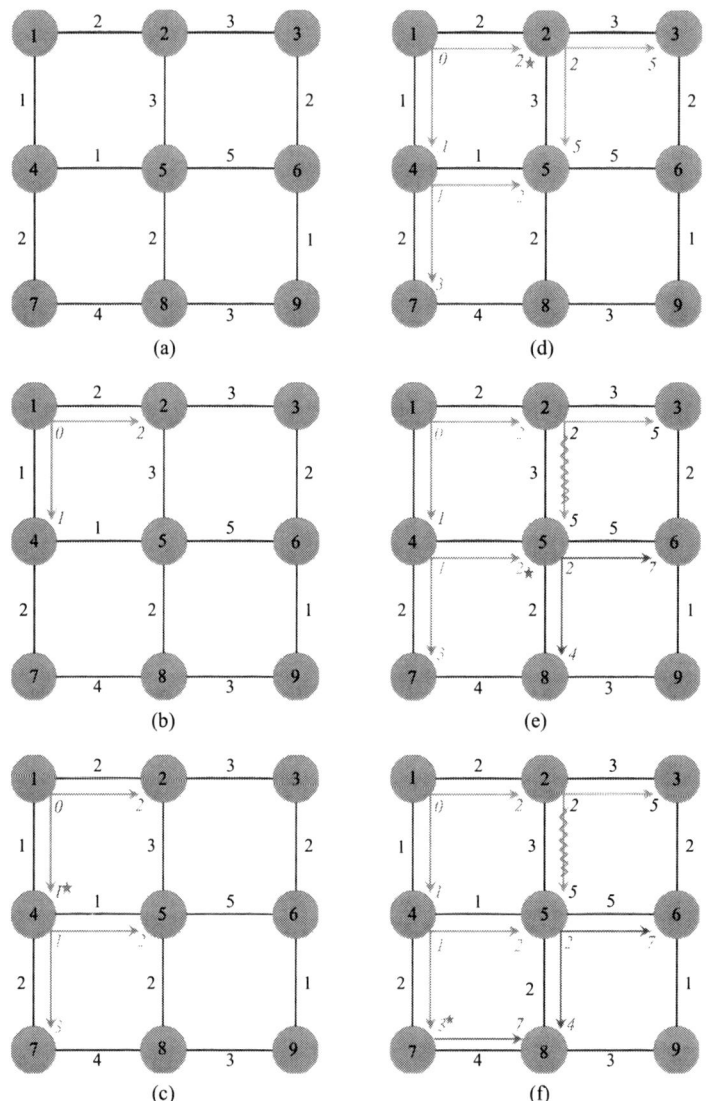

Figure 4.17. Moore's shortest path example.

then which mode of travel to use, and finally, which path to take. However, in reality, all these decisions are made simultaneously. For example, one might decide between shopping in the CBD using transit or in the suburb using automobile. By iterating the steps that include transportation impedance variables with revised values of these variables, the traditional approach can be modified. Thus, trip distribution, mode choice, and trip assignment can be applied using consistent estimates of level of service

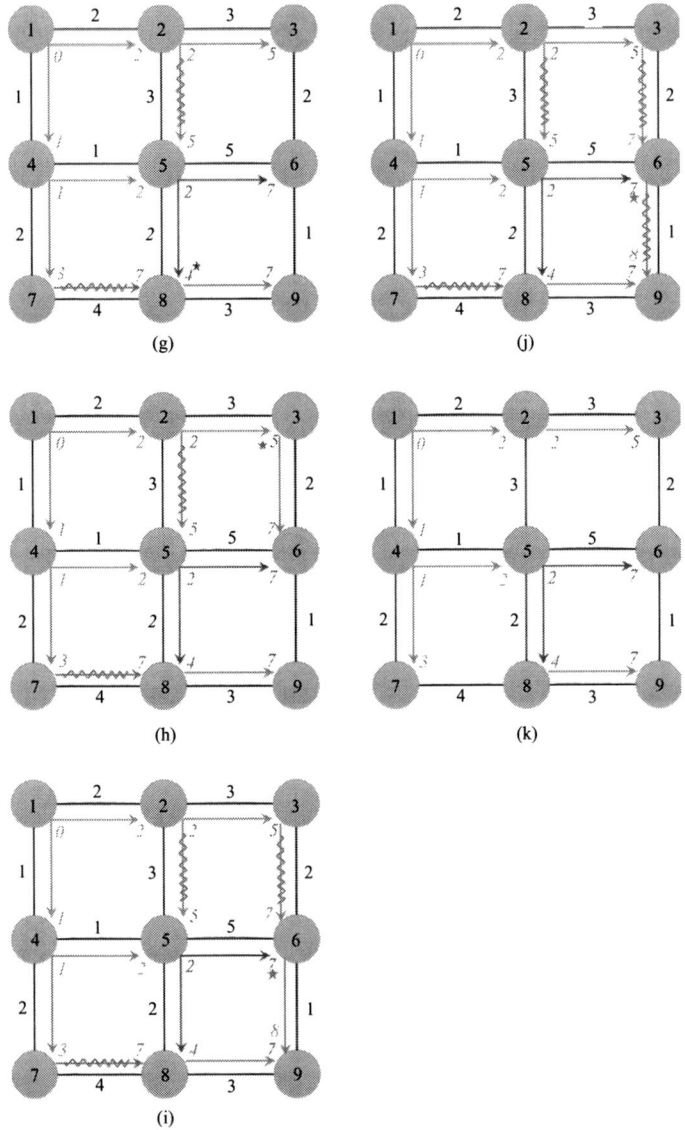

Figure 4.17. (Continued).

variables. These variables are feedback from trip assignment to the prior steps. The advantage of the feedback approach over the sequential process is that it provides more realistic trip-making forecasts and impedances between TAZ pairs than a simple sequential process. Almost all current four-step models use the feedback procedure. Figure 4.18 presents the feedback process.

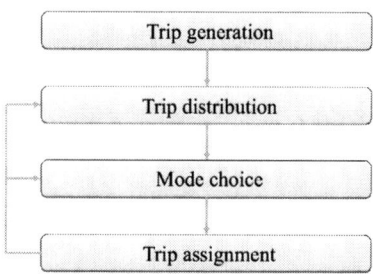

Figure 4.18. Feedback process.

Many modelers have experienced difficulty achieving convergence in the feedback process. Some practitioners use the feedback process only three or five times to avoid oscillation. However, others use the method of successive averages (MSA) proposed by Sheffi (1985). The average of link flow over all previous iterations with equal weight is used in MSA to guarantee convergence. Therefore, in the nth iteration, the link volumes of the current iteration as well as the $(n-1)$ previous iterations are weighted $(n-1)/n$. The iteration continues until it satisfies the convergence criteria (e.g., the difference between the volumes of the current and the previous iteration to be less than 5 percent). The link volumes resulting from this method are easily shown to converge for any pattern of highway assignments. The following steps are a summary of MSA procedure stated by Sheffi (1985, pp. 327).

> **Step 0**: Initialization: Perform a stochastic network loading based on a set of initial travel times $\{t_a^0\}$. This generates a set of link flows $\{x_a^1\}$. Set $n = 1$.
> **Step 1**: Update: Set $t_a^n = t_a(x_a^n)$, $\forall\, a$.
> **Step 2**: Direction finding: Perform a stochastic network loading procedure based on the current set of link travel times, $\{t_a^n\}$. This yields an auxiliary link flow pattern $\{y_a^n\}$.
> **Step 3**: Move: Find the new flow pattern by setting $x_a^{n+1} = x_a^n + (1/n)(y_a^n - x_a^n)$.
> **Step 4**: Convergence criterion: If convergence is attained, stop. If not, set $n = n + 1$ and go to step 1.

The feedback process can be simple or composite. Simple feedback adds highway level of service to the traditional sequential four-step model, while composite feedback combines highway and transit level of service to be used as impedance. Composite impedance requires computing composite variables for all zone pairs. The advantage of composite impedance is that trip distribution is based on both highway and transit characteristics, which is more realistic.

4.1.7 FOUR-STEP MODEL SHORTCOMINGS

Although the four-step model has been widely used for decades, it presents some significant drawbacks. It has been effective for previous transportation planning purposes that added new roads or increased road capacities. However, transportation planning policy has been changed in the last few decades to manage the existing transportation infrastructure, rather than add new infrastructure. Four-step models are not sensitive to these new policies.

In reality, many people do a chain of trips, while the four-step model does not consider trip chaining. It causes overestimation of some kinds of trips and underestimation of other kinds. Trip generation is not usually sensitive to changes in travel time; therefore, the effect of time-saving alternatives is not reflected.

The gravity model in the trip distribution step uses free-flow speed to compute travel times between zones, which are not realistic, especially in the congested networks. It also assumes constant trip lengths. These problems could be solved using feedback loops. Furthermore, it uses a single value for travel time, rather than using different travel times based on time of day. This problem can be partially tackled by having separate models for different times of day, usually morning peak, afternoon peak, mid-day, and off-peak. It is assumed that travel time factors by trip purpose remain constant over the years, which is an unrealistic assumption. More likely, the model overestimates near trips and underestimates far trips, because in the real world, people consider other factors than only travel time in choosing their destination. Trip distribution models are typically based on a single level of service variable, and link (highway) travel times have no effect on the number of trips predicted to alternative destinations. Furthermore, the gravity model is not behavioral. Distributions of trips by time of day are based on historical information. Therefore, changes in departure time and peak spreading are not reflected. Trip distribution can be replaced by a destination choice model to reflect travelers' behavior. Using logit models, utility of travelers is considered in the model.

The most problematic procedure in the four-step model is trip assignment. Traditional four-step models are static and do not consider time of day. They assign all travelers regardless of their departure time. But, in reality, travel rates vary by time of day. Modelers observed that trip rates vary in patterns that are consistently similar in all major urban areas. They assumed that the observed patterns for the base year will persist in the future and classified trips into three time periods: PM peak, AM peak, and off-peak periods. Such simplistic assumptions may no longer be adequate for large urban areas with significant traffic congestion. In the real world, increasing congestion in the peak period causes peak spreading. Peak spreading is changing the travelers' departure time to avoid long trips due to congestion, which is not accounted for in the four-step models. These models do not use trip departure times, and so, the only option of avoiding congestion in trip assignment is switching routes. Using dynamic traffic assignment and feedback process enables the model to do peak period factoring.

As mentioned earlier, trip assignment uses the link performance function to calculate link travel times. These functions are not sensitive to intersection delay, and assume delay occurs on the links. Thus, these models are not suitable for intersections, especially for signalized intersections. They assume that the travel time on a link is independent of flows on other links in the network, which is not true, particularly for the congested networks. They also do not consider queued vehicles in the traffic stream. Although these functions estimate the travel times well on non-congested links, they are unable to present a good estimation of travel time on links with volume greater than capacity. Using a traffic simulator could resolve the aforementioned problems.

Static assignment models assume that link flows and link trip times remain constant over the planning horizon. Thus, a matrix of steady-state OD trip rates are assigned to the network links. Although static equilibrium models are adequate for long-range planning analyses, they fail to capture the essential features of short-range planning analysis. Using dynamic traffic assignment could be an improvement, but it is impossible to be applied to the four-step model because this model is time-invariant and does not have the structure to keep trip departure times.

Another limitation of the traditional four-step model is that it assigns different modes of travel (e.g., transit and highway) separately, while, in reality, some modes share the road with each other (McNally 2007). A solution to this problem is using a composite impedance of different modes in the feedback loop.

4.2 ACTIVITY-BASED MODELS

The four-step model is a trip-based approach that focuses on trips without considering the reason for the trip, while activity-based models are based on the fact that people travel to pursue an activity. This enables researchers to better understand the behavioral basis of travelers' decision in mode choice, destination choice, route choice, and so on. Unlike trip-based models, activity-based models are tour-based and consider a logical trip chaining. Activity-based modeling dates back to Chapin (1971 and 1974) who studied human activity patterns in urban areas, and Hagerstrand (1970) who examined activity patterns in a time-space domain and specified series of constraints. The constraints include time-space constraint, biological needs (e.g., sleep) and resources (e.g., income), and inter-agent interactions (e.g., children).

Individuals have 24 hours and they decide how to allocate it among different activities. Figure 4.19 presents a daily activity itinerary. Some activities such as work and sleep occur at a specific fixed location, while others, such as shopping, may occur in different locations. The person wakes up at 6:30 AM, leaves home at 7:15 AM, and drops his or her kid to school at 7:30 AM. Then he or she gets to work at 8:00 AM. He or she leaves work for lunch at 12:30 PM, goes to post office at 1:00 PM, and goes back to work at 1:15 PM. He or she leaves work at 5:15 PM, goes grocery shopping at 5:45 PM, and gets home at 6:15 PM. He or she stays home until the next morning. All other activities such as eating dinner, watching TV, and sleeping happen at home.

Activity-based models are very useful in evaluating transportation policies, such as congestion pricing, and tolling strategies, such as HOV lanes, and HOT lanes, which is not possible in the aggregate trip-based approach in the four-step models.

Much activity-based travel demand modeling has been developed based on the utility maximization theory, that individuals make their activity-travel decisions to maximize their utility. Some others use computational process models (CPM), which is a set of rules in the form of if–then pairs. Agent-based modeling has also been used by some researchers.

Activity-based models generally yield a list of activities and trips, including detailed information about departure time, destination, and mode for each trip. This detailed output can be used in a dynamic traffic assignment model or can be aggregated into OD matrixes to be used in a static traffic assignment model.

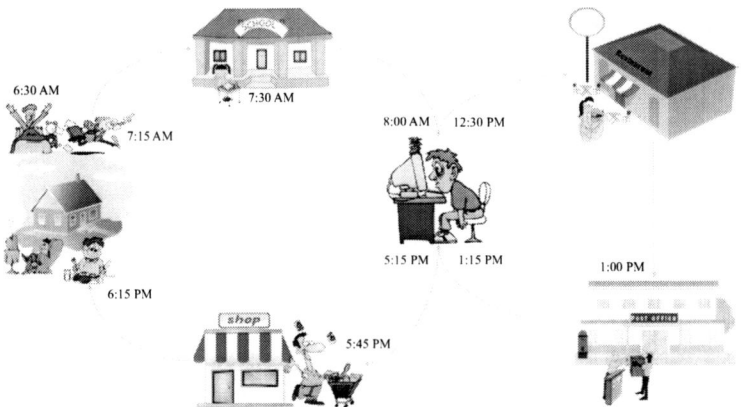

Figure 4.19. A sample of daily activity itinerary.

TRANSIMS is the first complete agent-based or activity-based package developed by Los Alamos National Laboratory under the Transportation Model Improvement Program (TMIP). It was funded by the U.S. Department of Transportation, Environmental Protection Agency (EPA), and Department of Energy. TRANSIMS is an integrated travel demand activity or agent-based model for large metropolitan areas. TRANSIMS was designed to replace the four-step travel demand model and address current issues such as sustainability, environmental impact, and emerging intelligent systems. It can also be used for Homeland Security modeling applications. TRANSIMS traces the movement of each individual in accomplishing their daily activities over the 24-hour period and simulates those movements over a detailed network. This agent-based microsimulation makes TRANSIMS a powerful tool to be used in the aforementioned applications. Some examples of those applications are as follows.

TRANSIMS enables emergency responders to evaluate regional emergency response plans based on their feasibility. It is capable of simulating the complex and highly dynamic effects of evacuation (Park 2011). Sadek (2012) developed models in TRANSIMS to quantify the impact of inclement weather on freeway traffic speed and showed that TRANSIMS can be used for online management of transportation systems. Samba and Park (2010) evaluated different safety treatments using TRANSIMS. Lee and Hobeika (2007) estimated drivers' response to a toll on a dynamic high-occupancy toll (HOT) lane in TRANSIMS.

4.3 AGENT-BASED MODELS

Agent-based modeling (ABM) or individual-base modeling is a new modeling paradigm, which is used in physics, biology, economics, social sciences, and transportation. It models the dynamic interaction between agents. ABM is an extension to cellular automation (CA) models (Wolfarm 1994). An advantage of ABM over CA is that the interaction between agents and between agents and their environment is asynchronous. CAs simultaneously make actions at constant time-steps, while ABM agents follow a sequential schedule of interactions in a discrete time-steps.

ABM is a microscopic bottom–up modeling, while most of other models (e.g., four-step) are macroscopic top–down modeling approaches. An agent has the following characteristics:

- An agent is a discrete individual with a set of characteristics.
- An agent interacts with other agents based on communication protocols and rules.

- An agent has some goals to achieve.
- An agent is autonomous and self-directed.
- An agent has the ability to learn and evolve its behavior over time based on its experience.

ABM is best applied to situations in which:

- the interactions between the agents are complex, non-linear, discontinuous, or discrete;
- the agents have complex behavior such as learning;
- each individual agent is potentially different; and
- the agents move and their positions are not fixed.

In the transportation field, ABM is appropriate for modeling systems in which human decisions and actions are a critical component, such as demand modeling.

Two of the well-known agent-based models for traffic simulation are TRANSIMS (https://fhwa.dot.gov/planning/tmip/resources/transims/) and MATSIM (http://matsim.org/). TRANSIMS was explained in the previous section. MATSIM is a large-scale agent-based traffic simulator that was developed after TRANSIMS. There are some similarities and some differences between TRANSIMS and MATSIM. They both are data-intensive, but MATSIM is faster than TRANSIMS and its traffic flow simulation is more simplified.

Agent-based models are well suited for microscopic simulation because they are disaggregated and can account for human behavior.

BIBLIOGRAPHY

Akçelik, R. 1991. "Travel Time Functions for Transport Planning Purposes: Davidson's Function, Its Time-Dependent Form and an Alternative Travel Time." *Australian Road Research* 21, no. 3, pp. 49–59, minor revisions in 2000.

Chapin, F.S. 1974. *Human Activity Patterns in the City*. New York: John Wiley.

Chapin, F.S., Jr. 1971. "Free-Time Activities and the Quality of Urban Life." *Journal of the American Institute of Planners* 37, pp. 411–417.

Hägerstrand, T. 1970. "What About People in Regional Science?." *Papers and Proceedings of the Regional Science Association* 24, pp. 7–24.

Hobeika, A.G. 2003. *Land Use and Transportation Planning, Power Point Slides*. Blacksburg, VA: Department of Civil and Environmental Engineering, Virginia Tech.

Lee, K.S., and A.G. Hobeika. 2007. "Application of Dynamic Value Pricing Through Enhancements to TRANSIMS." *Transportation Research Record*, 2003, pp. 7–16.

Lee, K.S., J.K. Eom, and D. Moon. 2014. "Applications of TRANSIMS in Transportation: A Literature Review." *Procedia Computer Science* 32, pp. 769–773.

Macal, C.M., and M. North. 2005. "Tutorial on Agent-Based Modeling and Simulation." *Proceedings of the 2005 Winter Simulation Conference.*

McNally, M. 2007. The Four Step Model. *Handbook of Transportation Modeling.* Amsterdam, Netherlands: Elsevier Science.

Meyer, M., and E. Miller. 2001. *Urban Transportation Planning: A Decision-Oriented Approach.* 2nd ed. New York, NY: McGraw-Hill.

Ortuzar, J., and L. Willumsen. 2005. *Modelling Transport.* 3rd ed. Hoboken, NJ: John Wiley & Sons, Ltd.

Patriksson, M. 2015. *The Traffic Assignment Problems—Models and Methods.* Mineola, NY: Dover Publications.

Sadek, A. 2012. "Using TRANSIMS for On-Line Transportation System Management During Emergencies." *FHWA Peer Exchange Meeting*, University at Buffalo.

Samba, D., and B.K. Park. 2010. "Development of the TRANSIMS Safety Evaluation Module and its Application on Large Truck Safety Treatments. *Research Report UVACTS-16-0154*, University of Virginia.

Sheffi, Y. 1985. *Urban Transportation Networks: Equilibrium Analysis with Mathematical Programming Methods.* Upper Saddle River, NJ: Prentice-Hall.

Spiess, H. 1990. "Technical Note—Conical Volume-Delay Functions." *Transportation Science* 24, no. 2, pp. 153–158.

Transportation Research Circular 2007. *Artificial Intelligence in Transportation: Information for Application*, E-C113.

Transportation Research Circular 2011. *Dynamic Traffic Assignment A Primer*, E-C153.

Wolfram, S. 1994. *Cellular Automata and Complexity.* Boston, MA: Addison Wesley.

CHAPTER 5

REAL-TIME SYSTEMS

Since the late 1980s, information has influenced travelers' route choice process; therefore, it became an essential component of traffic analysis and route choice models. Advanced traveler information systems (ATIS) were introduced in the 1990s to expand travelers' choice sets and let them make more informed route decisions. The dynamic traffic assignment (DTA), which was discussed in Chapter 4, emulates driver's decision-making processes by enhancing intelligent transportation systems (ITS) technologies. Traveler information can be personalized, such as in-vehicle information systems (IVIS), or non-personalized, which presents generic information to a wide range of road users.

In order to manage traffic, transportation system operators have gained access to real-time traffic data either directly or through third-party data providers, such as Inrix and Here. Traffic data is gathered by means of sensors, detectors, cameras, probe vehicles, or other technologies for a variety of purposes by traffic management centers (TMCs). The centers can then offer useful real-time travel information to travelers under pre-trip and en route information categories. The most relevant traffic information being communicated with motorists are:

- Estimated travel time to a known destination (with or without travel distance);
- Road work activities and incident awareness (with or without delay information); and
- Route guidance and diversion recommendation.

Figure 5.1 illustrates live traffic status surrounding the Washington, DC, network, along with the locations of cameras and dynamic message signs (DMS). This information can be conveyed to target travelers through DMS, mobile apps, live traffic channels, statewide 511 phone systems, and so on. As a widely utilized dissemination mechanism of ATIS, DMS

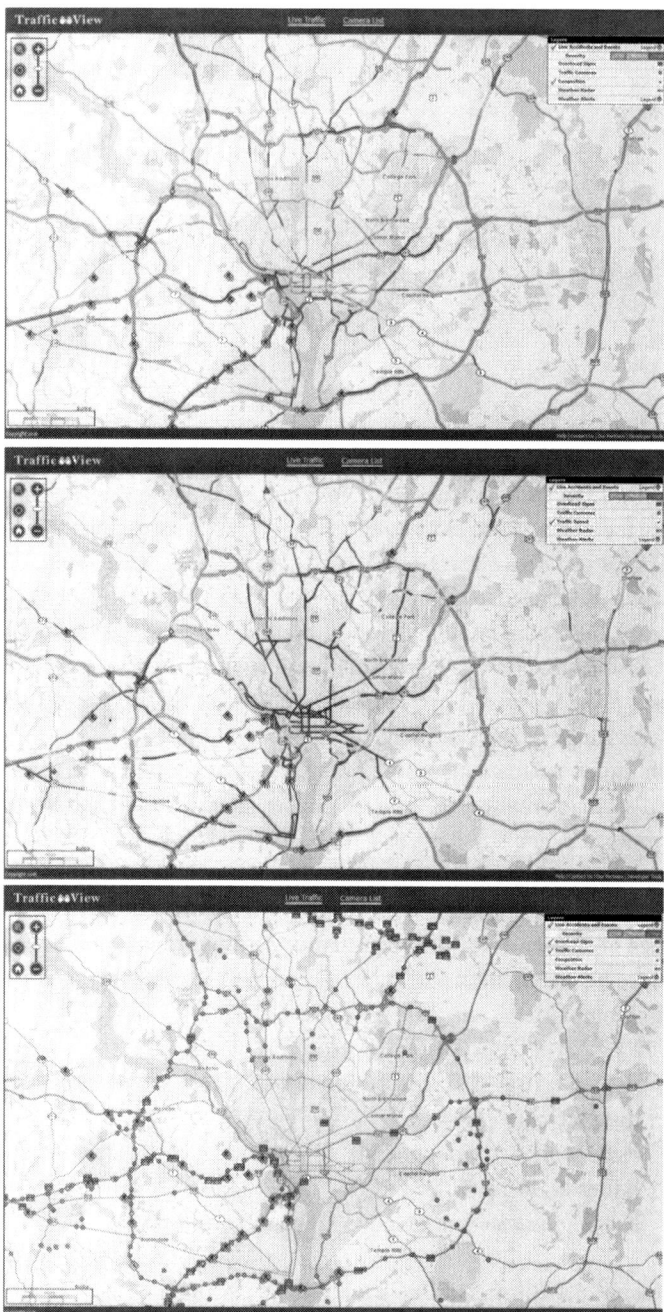

Figure 5.1. Live traffic in Washington, DC, area (top: congestion level, center: speed, bottom: DMS and camera locations).

Source: http://i95coalition.org/trafficview/

Figure 5.2. DMS as a traffic management tool (interstate 95 northbound toward Washington, DC).

are proven to be clear sources of travel information. DMS provide quick information to motorists en route about adverse road conditions, traffic incidents, travel time, speed control, managed lanes, traffic regulations, and road construction, among other safety and guide messages. The original postulation was that providing more traffic information in an effective way leads to better route choices by motorists, and field and simulator data have proven an acceptable level of compliance with the information provided. Figure 5.2 shows an example of usage of DMS to have the travelers make more informed decisions.

The Manual on Uniform Traffic Control Devices (MUTCD 2009) defines changeable message signs (CMS), as "traffic control devices capable of displaying one or more alternative messages." States may have complementary description of DMS; for instance, according to the Florida Department of Transportation, the purposes of the DMS are providing information on a change in current or future traffic conditions and also informing motorists to change travel speeds, change lanes, and divert to an alternate route. The effectiveness of DMS is strongly correlated with the quantity of information provided, such as a suggested detour, expected delay, and cause of delay. Compliance with route advice appeared to be limited and is strongly associated with drivers' personal interests, as well as information accuracy. Driver's age, driving years, annual mileages, income, driving style (risk-based, steady, and conservative), occupation, route familiarity, route choice style, perception of information, DMS content, and degree of trust in DMS, and even driver's mood are proven to be significant factors to explain DMS guidance compliance behavior. Driving simulator-based studies have found that appending delay time of the main route and travel time of the alternative route to DMS would achieve higher compliance rate.

5.1 TMC OPERATIONS AND FUNCTIONS

NCHRP 270 (1998) summarized TMC roles as gathering, synthesis, and dissemination of traffic-related data. According to the Federal Highway Administration (FHWA) (2004), four standard functions of a TMC are monitoring, managing events, providing services, and maintenance. TMCs are expected to carry out diverse functions, including traffic signal system management, congestion management, failure management, incident management, special event management, and emergency management. While TMC functions are considered ongoing in nature, each contains several discrete tasks, with a distinct start and end, conducted by a TMC technician.

Typical day-to-day tasks that TMC operators are involved in include monitoring of CCTVs, travel time displays, real-time traffic volume data, traffic signal status, incident detection reports, radio broadcast information, and social media inputs, among others. Traffic signal system management, which appears to be one of the most essential functions, consists of the following tasks: monitoring signal timing plans, verifying communication and detector status, identification of signal malfunctioning, and collaborating with signal timing engineers to adjust signal timing or phasing changes or implement a special timing plan due to a non-recurring condition.

Traffic incident management includes using all potential resources to reduce the duration and impact of incidents in a coordinated, systematic way to return the roadway back to normal traffic conditions as quickly as possible. This not only improves the safety of travelers involved in the incident and responders, but also reduces secondary crash risk by reducing incident recovery time. TMCs are expected to detect any sign of incident from all accessible sources (CCTVs, radio information, Bluetooth sensors, traffic detectors, and so on) and be ready to activate ITS devices, dispatch the appropriate team to the scene, notify the public, determine possible alternate routes, guide the traffic to alternates, and eventually, perform post-incident treatment.

TMCs may develop software tools to make short-term traffic prediction and develop an operator decision support system. When an incident occurs, it is expected to be quickly detected by TMC operators to respond accordingly. Figure 5.3 depicts Baltimore, Maryland's TMC display screens.

Figure 5.3. TMC displays in Baltimore, Maryland.

5.2 DEMAND MANAGEMENT THROUGH TRAFFIC INFORMATION

Transportation officials also employ traveler information as a powerful tool to manage travel demand. This can be an inexpensive method of preventing congestion or relieving existing congestion faster if travelers demonstrate a high compliance rate with the information and recommendation dispensed. Due to the emergence of new traveler information systems, demand management strategies have evolved from the traditional methods of pre-trip management, that is, changing the travel mode, work hours, and decision to telecommute. While most of these shifts were long-term and static decisions, newer demand management strategies focus on short-term, timely and dynamic changes to travel components, such as route and departure time.

5.3 ADVANTAGES AND LIMITATIONS

In a steady state network, average values of traffic attributes can be representative of traffic conditions. However, many characteristics used in transportation analysis are inherently random variables, including traffic volume and travel speed, that vary due to the stochastic nature of travel demand and human behavior. A dynamic modeling approach can better process the variability of random parameters.

BIBLIOGRAPHY

Divekar, G., H. Mehranian, M.R.E. Romoser, J.W. Muttart, P. Garder, J. Collura, and D.L. Fisher. 2011. "Predicting Route Choices of Drivers Given Categorical and Numerical Information on Delays Ahead: Effects of Age, Experience, and Prior Knowledge." *Transportation Research Record, Journal of Transportation Research Board* 2248, pp. 104–110.

FHWA, TMC Operator Requirements and Position Descriptions. 2004. Draft Report.

Kerkman, K., T. Arentze, A. Borgers, and A. Kemperman. 2012. "Car Drivers' Compliance with Route Advice and Willingness to Choose Socially Desirable Routes." *Transportation Research Record, Journal of Transportation Research Board* 2322, pp. 102–109.

Montes, C. 2008. "Guidelines for the Use of Dynamic Message Signs on the Florida State Highway System." Prepared for: Florida DOT, Traffic Engineering and Operations Office, ITS Section.

MUTCD 2009. "The Manual on Uniform Traffic Control Devices." 2009 Edition, Federal Highway Administration, U.S. DOT.

NCHRP Synthesis 270, Transportation Management Center Functions. 1998. Washington, D.C.: Transportation Research Board, National Academy Press.

Zhong, S., L. Zhou, S. Ma, and N. Jia. 2012. "Effects of Different Factors on Drivers' Guidance Compliance Behaviors under Road Condition Information Shown on VMS." *Transportation Research Part A* 46, no. 9, pp. 1490–1505.

CHAPTER 6

CALIBRATION AND VALIDATION TECHNIQUES

6.1 CALIBRATION VERSUS VALIDATION

Model calibration is a general term for estimating or adjusting model parameters to improve its practicality and predictability. Model calibration includes a range of techniques, some standard and some creative, to accurately reproduce observed or actual values. Model validation predicts the values of interest that are already known for another time period (e.g., another year) or another similar location (e.g., a comparable corridor) using the calibrated model. If the calibrated model predictions are close enough to the observed data, then the model is validated; otherwise, the model parameters and assumptions need to be revisited. Therefore, the modeler may return to the calibration step several times before an acceptable validation is achieved. These two concepts sometimes are interchangeably used; however, they are different and must be applied correctly. Historical data is the key element of calibration and validation.

There are some global measures of effectiveness (MOEs) to quantify how well a traffic or transportation model is performing its job, whether it is forecasting traffic conditions in a future scenario or gauging the importance of different socio-economic factors on a travel choice problem. MOEs are sets of criteria that quantify a model's performance and provide a measuring tool for the modeler to evaluate a model's goodness of fit. There are normally field-measured values for the MOEs to provide a comparison basis for the calibration steps. Any efforts toward improving MOEs are part of the calibration process. Calibration can be as simple as fixing nodes and links connection and geometry fixations to manipulating driver's behavior, including speed choice, acceleration or deceleration behavior, overtaking behavior, lane changing, gap acceptance, look ahead distance, compliance level with travel information, and so on. Calibration also entails tweaking the distributions of all these aforementioned parameters of the driver population because drivers are heterogeneous and traffic has a stochastic nature.

For instance, queue length and signal delay are two major MOEs in modeling a traffic signal, whereas model prediction of these MOEs is expected to fall within an acceptable range of the observed values.

Practitioners and modelers follow the national and local guidelines on calibration precision and thresholds. In the following sections, relevant guidelines and standards will be presented.

The coefficient of determination (R^2) measures how properly observed outcomes are replicated by the statistical model, by indicating how much of the total variation in outcome variable is explained by the model. It generally ranges from zero to one, with one being the best fit. While R^2 is a widely used measure of model performance in multiple regression models, there is no clear analog to R^2 as a measure of goodness-of-fit for logit and probit models. Several substitutes to the standard R^2 have been proposed by statisticians to deal with categorical outcomes' dependent variables. An entropy-based R^2 indicator, also known as pseudo-R^2, is one of the most popular indicators in social sciences. However, unlike the traditional R^2, pseudo-R^2 is not interpretable based on the variance of logistic regression outcome. Nagelkerke (1991) generalized the definition of coefficient of determination to more general models as shown in Equation 7.8. This equation is consistent with the maximum likelihood function to estimate model parameters.

$$R^2 = 1 - \left(\frac{L(0)}{L(\theta)} \right)^{2/n} \qquad \text{(Equation 7.8)}$$

where,

$L(0)$: Maximum likelihood for the model without any predictor; and
$L(\theta)$: Maximum likelihood for the final model with all predictors.

6.2 CALIBRATION AND VALIDATION IN TRAFFIC SIMULATION MODELS

Traffic simulation is a widely used modeling technique in transportation engineering that can be complementary to transportation planning models and provide another useful perspective to decision makers. Traffic simulation is not essentially an alternative to planning models; however, it can take advantage of planning model outputs, develop multiple scenarios,

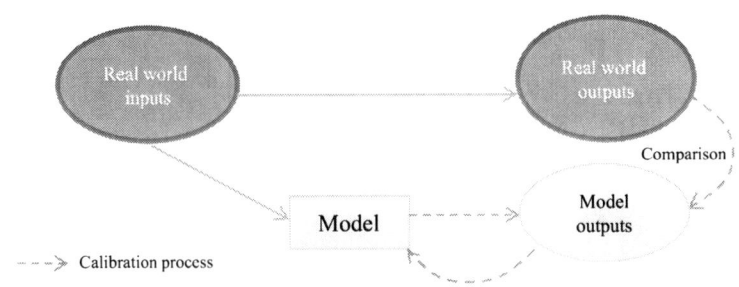

Figure 6.1. Model calibration process.

and effectively evaluate alternatives' performance with more detailed inputs. Traffic simulation normally concentrates on a shorter horizon, smaller scale network, and narrower changes to the transportation supply or demand. It can answer operational questions and visualize the consequences of any change in the geometry, traffic control systems, and ITS domain.

The calibration of simulation models is a key step in developing a reliable model to guide major policies based on the model's output. The calibration of model parameters includes adjusting default parameters with local and observed driving patterns. Driving behaviors and vehicle characteristics can be unique to any geographic area, although these data are tricky and costly to be collected. All traffic simulation software packages contain default values for behavioral parameters that are either governed by standards and guidelines or most fit the region for which traffic data has been collected and models are calibrated. Fine-tuning network details is necessary to closely replicate field values. Figure 6.1 shows the calibration process of a model that concludes model outputs within an acceptable error range compared to the real-world outputs. It is noteworthy that many calibration steps should be supported by engineering knowledge of the study site and modeling experience.

6.2.1 MOEs AND DATA COLLECTION

MOEs are selected according to the purpose and context of the study. It is important to note that MOEs' establishment, data collection techniques, and calibration strategies are all contingent upon the level and type of the model. Aggregate models, such as trip-based travel demand models (discussed in Chapter 4) and macroscopic traffic models, are coded and calibrated differently than disaggregate models, such as agent-based or traffic microsimulation models. It is vital to collect sufficient data inputs

Table 6.1. Data input requirements for model calibration

Traffic volumes	Mainline throughputs or corridor counts Turning movements at intersections and interchanges
Speed and travel time	Space mean speed
	Time mean speed
	Point-to-point travel time
	Intersection delay
Queuing* and density	Queue length at signalized intersection per lane group
	Queue length at roundabouts
	Queue length at merging points
	Density of freeway sections
Transit data	Frequency or schedule
	Transit travel time
	Ridership data (boarding or alighting)

* Depending on the software tool used and the agency's specific guideline, simulated queue might indicate the average, maximum, or the 95th percentile queue; however, it is common to be reported in unit of length (i.e., feet). Hence, it is recommended to measure field queue in a unit of distance and not number of vehicles. If number of vehicles is counted, they should be converted to length using appropriate vehicle composition.

of the existing conditions to facilitate model calibration. Table 6.1 lists the key input variables to be measured in the field for a basic calibration. These variables are typical outputs from simulation runs in most simulation packages. Different local and state agencies may require different MOEs noted in their manuals or guidelines.

It is noteworthy that model input data that is essential for coding the model is different from calibration or validation data and must be collected to properly build the model. A list of required data to code a model is as follows:

- Network geometry and lane configuration: link distance, number of lanes, lane width, pocket lanes, length of turning bays, longitudinal grades, and so on.
- Traffic volumes and turning movement volumes for the peak period in 15-minute intervals.
- Traffic composition and heavy vehicle percentage: Most of the simulation software tools can accept heavy vehicle percentages

for each turning movement individually. Trucks and buses can significantly impact traffic performance, especially in intersections, weaving areas, and steep roadways, due to their lower acceleration and deceleration rates.

- Traffic signal parameters: signal control type (pretimed vs. actuated), cycle length, split times, offset (for coordinated signals), yellow and all red times, phase sequence, transit signal priority and preemption settings, pedestrian phases and minimum times, detectors setting, recall modes, and so on.

6.2.2 ERROR CHECKING AND REASONABLENESS

In this stage, model inputs mainly in the domain of geometry and vehicle inputs (volume and composition) will be verified to ensure the model is appropriately coded. It is assumed that the collected data has formerly undergone a thorough quality assurance process, which ensures data completeness and reasonableness, and good travel demand data is coded in the model. For instance, one should ensure that the demand data is collected under normal weather conditions, on a routine workday, and with no incident or roadwork affecting the study area. It is also presumed that traffic volumes along a corridor (an arterial with multiple intersections or a freeway with multiple ramps and loops) have been adequately balanced and no major sink or source is needed to be added to the model to justify any volume imbalances.

6.2.3 MODEL CALIBRATION

Simulation models require multiple calibration steps as follows. While stages A to C discuss the parameters that can be deterministic, stage D adopts stochastic variables that require adjustment of mean, variance, range, and distribution of some key operational and behavioral parameters that are specific to the study area.

- A. Network specification and geometry adjustment: Must ensure link length is greater than observed queue; all turning movements at an intersection are created; and so on.
- B. Traffic control setting adjustment: Ensure priority rules in conflicting points are properly set up, such as stop and yield signs, right turn on red (RTOR), conflicting pedestrians.

C. Animation check and reality control of traffic movement along the network.
D. Operational and behavioral variables' adjustment.
   - Travel time and speed: link speed, reduced speed areas, right- and left-turn speeds
   - Acceleration and deceleration rates and behavior
   - Lane changing, car following, and gap acceptance behavior
   - RTOR acceptance
   - Drivers' interaction, look-back and look-ahead distances
   - ATIS compliance

In this iterative process, driving behavior, vehicle and traffic flow characteristics are adjusted so that the model MOEs are in a reasonable range of field-measured MOEs. This involves refining various model parameters, requiring both field knowledge and expertise in the modeling tool utilized.

One major challenge in dealing with traffic models is the measurement of travel demand, as it may be misinterpreted as volume throughput. In most under-saturated traffic conditions, a deterministic approach will lead to a nearly accurate estimate of model MOEs when compared to field-measured values; however, in oversaturated conditions, where the capacity of a transportation facility falls short, there will be unserved demand for a period of time. Common examples of these capacity-restrained facilities are signalized or unsignalized intersections, reduced speed zones in suburban arterials, and freeway weaving sections. Such unmet demand will often appear in the form of queues backing up upstream of the choke point. Examples 6.1 and 6.2 demonstrate throughput versus demand and queuing computation for a cycle failure condition in a signalized intersection, respectively.

Example 6.1: The following traffic signal consists of three approaches: eastbound (EB), westbound (WB), and northbound (NB). While the one-way NB approach has one lane, the EB and WB approaches each carry two traveling lanes. Eastbound left (EBL) has an exclusive lane, leaving one lane for eastbound through (EBT) movement. Travel demands for the three approaches in six successive time intervals are shown in the following table. Assume intersection capacity of 1,000 vph for each traveling lane in any approach when the light is continuously green. Also, assume a three-phase 100-second cycle length traffic signal is controlling the intersection with the following effective green times: EBL: 20 sec., EBT: 70 sec., WBT: 50 sec., and NB: 30 sec. Disregarding the impact of turning movements on the saturation flow rate, calculate each lane group's throughput for each time interval.

Demand volumes per time interval for each approach

Time interval	EBL	EBT	WBT	NB
7:00–7:15	30	100	200	60
7:15–7:30	40	150	240	80
7:30–7:45	60	150	270	65
7:45–8:00	55	200	250	85
8:00–8:15	40	150	260	75
8:15–8:30	35	130	210	60

Solution 6.1: In order to estimate the throughput, we should first determine the actual capacity of each lane group considering the number of lanes and the amount of green time they receive. Here is the equation:

$$\text{Time Interval Capacity} = C_B \times f_{TI} \times f_g \times L.$$

where,

C_B: Base capacity (1,000 vphpl in this example)
f_{TI}: Time interval ratio (15 min. per one hour in this example)
f_g: Effective green time ratio (lane group's green interval divided by the cycle length)
L: Number of lanes

Thus, the capacity of each lane group for a 15-min. time interval is as follows:

Time interval capacity calculation process

Lane group	EBL	EBT	WBT	NB
C_B	1,000	1,000	1,000	1,000
f_{TI}	0.25	0.25	0.25	0.25
f_g	0.2	0.7	0.5	0.3
L	1	1	2	1
Time interval capacity	50	175	250	75

The following table shows the throughput for each lane group per time interval. As can be seen, throughput can never exceed lane group's capacity. The red cells indicate an overcapacity situation where the demand is higher than the capacity and queue is building due to unserved demand.

The green cells indicate unserved demand from a prior interval is being served in the following interval.

Throughput volumes per time interval for each approach

Time interval	EBL	EBT	WBT	NB
7:00–7:15	30	100	200	60
7:15–7:30	40	150	240	75
7:30–7:45	50	150	250	70
7:45–8:00	50	175	250	75
8:00–8:15	50	175	250	75
8:15–8:30	40	130	240	70

This fact demonstrates the subtle difference between the demand and throughput and serves as a caution to data collection strategies and findings. What can be collected in the field as turning movement volumes is throughput, rather than the actual demand. To derive demand, the approach's queue length must be collected along with the throughput to account for the unmet demand as well as the served demand.

Example 6.2: For the problem stated in Example 6.1, determine the intersection's longest queue length due to unmet demand, if the queue is being formed uniformly across adjacent lanes.

Solution 6.2: In comparison with the preceding two tables (throughput volumes and time interval), one can determine the following:

Cumulative queue at the end of each time interval for each approach (number of vehicles)

Time interval	EBL	EBT	WBT	NB
7:00–7:15	0	0	0	0
7:15–7:30	0	0	0	5
7:30–7:45	10	0	20	0
7:45–8:00	15	25	20	10
8:00–8:15	5	0	30	10
8:15–8:30	0	0	0	0

However, the WBT approach contains two lanes; therefore, the queue divides into half if equal traffic will be in each WBT lane. Thus, although the WBT approach earned the maximum cumulative queue (30 vehicles), the longest queue will be formed along the EBT approach (25 vehicles per lane).

Yet another stochasticity source of simulation models arises due to the inherent complexity of driving behavior and drivers' lane utilization. When drivers stop behind a traffic signal in a multiple lane arterial, drivers do not necessarily choose a driving lane to optimize the queue length. Although it is intuitive to stay in a lane with the shortest existing queue, there are many indications why drivers would not change lanes if they are already in a lane with a longer queue:

- Hesitation to change lanes due to safety, or simply it is too late to do so;
- Incorrect perception of queue length along different traveling lanes; and
- The existing lane affords them safer and faster access to the next decision point (i.e., a turning movement at a downstream traffic signal).

Therefore, the default parameter of lane utilization should be considered for possible adjustment with the support of field-observed data when calibrating a microsimulation model. Figure 6.2 illustrates the asymmetrical lane utilization between two left lanes at EB traffic, where the first left lane shows a queue of five vehicles, totaling 135 ft, while the far-left lane experienced a queue of nine vehicles (240 ft). Figure 6.3 refers to a similar behavior for two (WB) through lanes, one with four vehicles or 90-ft queue length, whereas the other lane accommodates 11 vehicles and a 280-ft-long queue.

Figure 6.2. Lane utilization example: Two EB left lanes.

Figure 6.3. Lane utilization example: Two WB through lanes.

Example 6.3: Determine the utilization factors for the lane groups denoted in Figures 6.2 and 6.3, if the condition shown is the prevalent situation.

Solution 6.3: The Highway Capacity Manual (HCM) defines the lane utilization factor (f_{LU}) as the ratio of the lane group's average volume per lane to the volume of the heaviest lane. Therefore, for the EBL approach in Figure 6.2, the $f_{LU} = 0.78$, which is $(5 + 9)/2/9$, and for the WBT approach in Figure 6.3, the $f_{LU} = 0.68$, which is $(4 + 11)/2/11$. This fact illustrates the imbalanced usage of adjacent lanes in one approach.

A typical calibration error occurs when throughput is coded in the model instead of actual demand. Contrary to Example 6.3, when throughput is collected in the field for a saturated condition, it should be converted to demand volume to fully replicate field conditions. Example 6.4 demonstrates a process to convert throughput volume to demand volume.

Example 6.4: Assume throughput and queue are collected at an intersection approach with two traveling lanes under an oversaturated condition. One crew member counted the throughput volume (vehicles that passed the intersection stop bar per time interval) and the other counted unserved demand during the same interval (vehicles remained in queue at the end of the interval). The following table shows throughputs and queues for a two-hour period. Determine the peak hour demand volume, assuming a lane utilization factor of 0.9.

Time interval	Throughput	Maximum observed queue (per lane)
7:00–7:15	105	0
7:15–7:30	130	2
7:30–7:45	123	13

7:45–8:00	126	21
8:00–8:15	125	24
8:15–8:30	124	20
8:30–8:45	120	10
8:45–9:00	116	0

Solution 6.4: The following table depicts the process to obtain demand volume per interval by first estimating the number of queued vehicles in the lane with minimum queue length as follows:

$$\text{2nd lane queue} = \text{roundup}\,(\text{1st lane queue} \times (2f_{LU} - 1))$$

The total queue includes the number of queued vehicles in both lanes. To calculate total demand, throughput is added to the additional queued vehicles formed by the end of each interval. For instance, although there were 38 queued vehicles by 8:00, 24 of those belonged to the previous interval demand and have been served during that interval.

Time interval	Second lane queue	Total queue	Added queue	Total demand
7:00–7:15	0	0	–	105
7:15–7:30	2	4	4	134
7:30–7:45	11	24	20	143
7:45–8:00	17	38	14	140
8:00–8:15	20	44	6	131
8:15–8:30	16	36	–8	116
8:30–8:45	8	18	–18	102
8:45–9:00	0	0	–18	98

Hourly demand (four successive 15-min. intervals) is calculated and shown in the following table. Hourly throughput is also shown for comparison. While throughput data showed a peak hour volume of 504 vph, the actual peak hour demand volume is 548 vph. Figure 6.4 compares these two variables. The shaded areas 1 and 2 in Figure 6.4 are the same and represent the unmet demand. It is noteworthy to emphasize that throughput volumes can hardly generate the observed queue length in modeling or simulation tools. Therefore, any attempt to calibrate factors other than volume, such as saturation flow rate, acceleration and deceleration

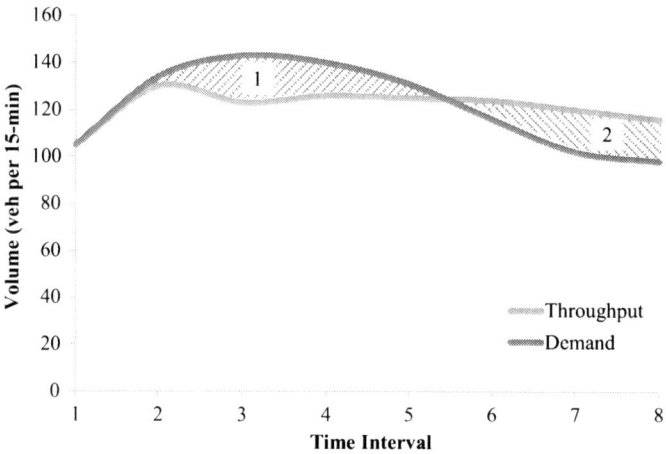

Figure 6.4. Throughput versus demand in Example 6.4.

coefficients, and minimum gap, to reproduce such queue may degrade the model's credibility and negatively impact its predictive power.

Hourly period	Hourly throughput	Hourly demand
7:00–8:00	484	522
7:15–8:15	504	548
7:30–8:30	498	530
7:45–8:45	495	489
8:00–9:00	485	447

6.2.3.1 Calibration Strategies, Quick Guide

A. Peak Hour Factor (PHF) or 15-min. peak volumes: Some analysis tools may require PHFs; however, microsimulation tools may need the actual 15-min. volumes for the peak period; therefore, they do not need a PHF factor.

B. Headway or gap: This behavioral parameter can determine the capacity of the lane group, and the user may need to override the software's default values.

C. Link speed and speed distribution: While macroscopic simulation tools are contingent on free-flow speed (FFS), microsimulation tools demand speed distribution of the vehicles traveling on a link as yet another implication of deterministic versus stochastic models. When intended to field measure the FFS, it must occur during the free-flow condition when the test vehicle's speed is not impacted by other vehicles in the roadway and all traffic control devices are in favor of the test vehicle. Florida Department of Transportation Recommends FFS to be set 5 mph above the posted speed limit when field data is not available.

    a. In microsimulation models, a distribution plot of vehicles' speeds, including minimum, maximum, mean, and median speed will be helpful to depict the real-world speed distribution.

D. Saturation flow rate: This value might be adjusted based on field knowledge. The HCM 2010 recommends 1,900 vphpl, which is the maximum throughput or capacity of a traveling lane when traffic flow is not impacted by adjacent traffic (parking maneuvers, transit stops, bike activities, and so on), geometry (gradient, narrow lanes, and so on), and other environmental conditions (pedestrian crossing, turning movements, and so on).

    a. Measuring the capacity is extremely important to determine whether a traffic condition is operating under or over capacity. Macrosimulation tools, such as HCS, Synchro, or SIDRA, are incapable of modeling oversaturated conditions (i.e., $v/C > 1$). Thus, a microsimulation tool, such as CORSIM, VISSIM, or SimTraffic, is required to model such conditions to generate performance MOEs.

E. Multiple runs: Microsimulation models generate traffic stream using random parameters, including volume per time interval, gap, vehicles' speed, lane changing behavior, and so on. It is recommended to simulate multiple runs that indicate various traffic conditions and report the average values of MOEs.

6.2.4 MODEL VALIDATION

Unlike calibration, model validation is a one-time assessment of the solidity of the model showing how practical the model is to predict a future scenario. Contingent upon the availability of dataset and data collection scope, the model's forecasting power is evaluated using a new dataset. There should be field-verified MOEs for these new datasets to be compared with the model's output to demonstrate the reliability of the model. The model is used with its calibrated parameters, and the parameters will not be adjusted as a result of validation. The same or different thresholds might be used in the calibration process when evaluating the model's MOEs against field-measured MOEs. It is not unlikely that a model fails to meet validation cutoffs when it has already passed calibration thresholds.

It is highly recommended that calibration or validation data be collected simultaneously with the project demand data (turning movement counts, throughput, speed, and so on). There might be significant differences in day-to-day traffic status, and the demand may vary, and consequently, the MOEs may vary to a larger extent. Let us assume a circumstance in which traffic volume data is collected on Day 1 to code the model as an input, and queue lengths are measured on Day 2 to calibrate the model as an MOE. While queue length cannot be a simple linear function of traffic volume, any attempt at calibrating the model by meeting the thresholds may be deemed impractical, and calibrated parameters may be misrepresentative or misleading.

6.3   CALIBRATION AND VALIDATION IN PLANNING

In planning, calibration is performed to make sure that the base model represents the real world.

The calibrated model for the base year is validated. Then, the calibrated and validated base model is used for the design years to predict future traffic flow. There are two validation methods, historical and backcasting. In the historical method, the current-year travel demand is forecasted using a prior-year model and the results are compared with the actual observed data in the current year. In the backcasting method, the prior-year travel demand is forecasted using a current-year model and the results are compared with the actual observed data in the previous year.

In a four-step model, calibration is performed in each step separately, as well as the overall calibration. Besides comparison with the observed data, the reasonableness of the output and parameters in each step is checked. The reasonableness check includes comparison of the model output and parameters with those of similar areas. Furthermore, sensitivity of the model to the input changes is investigated.

6.3.1   TRIP GENERATION CALIBRATION

The calibration of the trip generation step is estimating production and attraction sub-models parameters (coefficients) if a regression model is applied, as explained in Chapter 5.

A reasonableness check would be to determine the number of trips. The average number of trips in the United States is three to four per person and nine to ten per household. Another check is to examine traffic analysis zones (TAZs) with very high and low productions and attractions, as well as comparing estimated trips with the observed ones (from the household travel survey) in the following categories: the whole study area, trip percentages by trip purpose, and the number of trips in different income categories. If the compared trips are close enough, the model is calibrated; otherwise, revisions must be performed. Revisions can be done in the land use model to estimate a more realistic land use file to be used in trip generation. Another revision can be modifying production and attraction sub-models.

6.3.2 TRIP DISTRIBUTION CALIBRATION

Calibration of the gravity model estimates travel time factors (F_{ij}) and zone-to-zone adjustment factors (K_{ij}) for the base year, which is explained next.

The number of trips distributed between each TAZ pair is acquired from the household travel survey. As the survey does not include the whole population in the area, these numbers are expanded to represent all trips distributed in the study area. The expanded trip table (T_{ij}) is used to estimate F_{ij} and K_{ij} in Equation 4.3.

As no direct estimation procedure is available that would compute F_{ij} and K_{ij} factors from the known T_{ij}, P_i, and A_j, K_{ij} is assumed to be one (for all $i, j = 1, 2, ..., n$) and then, a trial-and-error solution is used to determine the values of F_{ij}, subject to the following conditions.

$$\sum_{i=1}^{n} P_i = \sum_{j=1}^{n} A_j$$

$$\sum_{j} T_{ij} = P_i \qquad \text{(Equation 6.1)}$$

$$\sum_{i} T_{ij} = A_j$$

6.3.2.1 Trial and Error Procedure to find F_{ij}

1. Group zones within five-min time intervals (or other time intervals of your choice) and assume an initial value (usually 1) for F_{ij}.
2. Calculate T_{ij} values using the formula in Equation 4.3.
3. Compare the number of trips obtained by the formula and the observed values in the base year for each of the time interval zone groups.

4. Adjust the travel time factors for each time interval as follows:

$$F_{ij_n} = F_{ij_{n-1}} \frac{\text{Total trips observed from data}}{\text{Total trips calculated from Gravity Model}}$$

5. If $F_{ij_n} / F_{ij_{n-1}}$ is close to 1 (e.g., between 0.975 and 1.025), stop; otherwise, return to step 2.

The following small example presents the detailed procedure to find F_{ij} for a 3-TAZ study area. Travel time and number of trips between TAZ pairs are given in Table 6.2.

The first step is to group TAZs for travel time of 0–5 minutes, 5–10 minutes, and 10–15 minutes. TAZ pairs of 1–1, 2–2, and 3–3 have a travel time of 0–5 min., while TAZ pairs of 1–3 and 3–1 have a travel time of 5–10 min., and TAZ pairs of 1–2, 2–1, 2–3, and 3–2 have a travel time of 10–15 min. The total number of trips in the first group is calculated based on the given trip interchange matrix as presented in Table 6.3.

The second step is to calculate T_{ij} values as follows:

$$T_{11} = \frac{300 \times 200}{550} = 109, \ T_{12} = \frac{300 \times 160}{550} = 87, \ T_{13} = \frac{300 \times 190}{550} = 104$$

$$T_{21} = \frac{100 \times 200}{550} = 36, \ T_{22} = \frac{100 \times 160}{550} = 29, \ T_{23} = \frac{100 \times 190}{550} = 35$$

Table 6.2. Travel time and trip table for the 3-TAZ example

Travel time (min.) for base year			
F/T	1	2	3
1	2	12	8
2	11	5	14
3	6	15	3
Trip table for base year			
F/T	1	2	3	P_i
1	40	110	150	300
2	50	20	30	100
3	110	30	10	150
A_i	200	160	190	550

Table 6.3. Trial-and-error procedure to find F_{ij} for the 3-TAZ example (first trial)

Group	Travel time interval (min.)	TAZ	Total observed trips	F_{ij_1}	Total calculated trips	F_{ij_2}
1	0.1–5.0	1–1, 2–2, 3–3	$40 + 20 + 10 = 70$	1	$109 + 29 + 52 = 190$	$\dfrac{70}{190} \times 1 = 0.368$
2	5.1–10.0	1–3, 3–1	$150 + 110 = 260$	1	$104 + 55 = 159$	$\dfrac{260}{159} \times 1 = 1.644$
3	10.1–15.0	1–2, 2–1, 2–3, 3–2	$110 + 50 + 30 + 30 = 220$	1	$87 + 36 + 35 + 44 = 202$	$\dfrac{220}{202} \times 1 = 1.090$

$$T_{31} = \frac{150 \times 200}{550} = 55, \; T_{32} = \frac{150 \times 160}{550} = 44, \; T_{33} = \frac{150 \times 190}{550} = 52$$

Then, calculate the summation of trips in each group and find adjusted F_{ij} for each group for the next iteration. The F_{ij} values to be utilized in the next iteration are presented in Table 6.4.

In the second iteration, using the calculated F_{ij} factors from the first trial, find the new T_{ij} values using Equation 4.3 as follows and update F_{ij} as presented in Table 6.5.

$$T_{11} = \frac{300 \times 200 \times 0.368}{200 \times 0.368 + 160 \times 1.090 + 190 \times 1.644} = 39,$$

$$T_{12} = \frac{300 \times 160 \times 1.090}{200 \times 0.368 + 160 \times 1.090 + 190 \times 1.644} = 93,$$

$$T_{13} = \frac{300 \times 190 \times 1.644}{200 \times 0.368 + 160 \times 1.090 + 190 \times 1.644} = 167,$$

$$T_{21} = \frac{100 \times 200 \times 1.090}{200 \times 1.090 + 160 \times 0.368 + 190 \times 1.090} = 45,$$

$$T_{22} = \frac{100 \times 160 \times 0.368}{200 \times 1.090 + 160 \times 0.368 + 190 \times 1.090} = 12,$$

$$T_{23} = \frac{100 \times 190 \times 1.09}{200 \times 1.090 + 160 \times 0.368 + 190 \times 1.090} = 43,$$

$$T_{31} = \frac{150 \times 200 \times 1.644}{200 \times 1.644 + 160 \times 1.090 + 190 \times 0.368} = 86,$$

$$T_{32} = \frac{150 \times 160 \times 1.090}{200 \times 1.644 + 160 \times 1.090 + 190 \times 0.368} = 46,$$

$$T_{33} = \frac{150 \times 190 \times 0.368}{200 \times 1.644 + 160 \times 1.090 + 190 \times 0.368} = 18$$

Table 6.4. F_{ij} matrix for the 3-TAZ example after the first trial

F_{ij} matrix (after first trial)			
F/T	1	2	3
1	0.368	1.090	1.644
2	1.090	0.368	1.090
3	1.644	1.090	0.368

Table 6.5. Trial-and-error procedure to find F_{ij} for the 3-TAZ example (second trial)

Group	Travel time interval (min.)	TAZ	Total observed trips	F_{ij_2}	Total calculated trips	F_{ij_3}
1	0.1–5.0	1–1, 2–2, 3–3	40 + 20 + 10 = 70	0.368	39 + 12 + 18 = 70*	$\frac{70}{70} \times 0.368 = 0.369$
2	5.1–10.0	1–3, 3–1	150 + 110 = 260	1.644	167 + 86 = 253	$\frac{260}{253} \times 1.644 = 1.688$
3	10.1–15.0	1–2, 2–1, 2–3, 3–2	110 + 50 + 30 + 30 = 220	1.090	93 + 45 + 43 + 46 = 227	$\frac{220}{227} \times 1.090 = 1.057$

*Trips are rounded but the summation without rounding is almost 70 rather than 69.

Because F_{ij_3}/F_{ij_2} values are between 0.975 and 1.025, we stop the iteration process. It means that the calculated trips are within 0.975 and 1.025 of the observed trips. The trip table is presented in Table 6.6. Travel time factors (F_{ij} s) should be plotted against travel time and a graph similar to Figure 6.5 must be produced. If they do not follow the graph, it will be very difficult to balance calculated trip attractions and productions with the observed ones.

6.3.2.2 Row and Column Factoring

Because total attracted trips to TAZs are different from the original one, successive column and row factoring is needed to re-estimate zonal productions and attractions to obtain the same total attractions

Table 6.6. Updated trip table for the base year in the 3-TAZ example

Updated trip table for the base year				
F/T	1	2	3	P_i
1	39	93	167	300
2	45	12	43	100
3	86	46	18	150
A_j	171	151	228	550

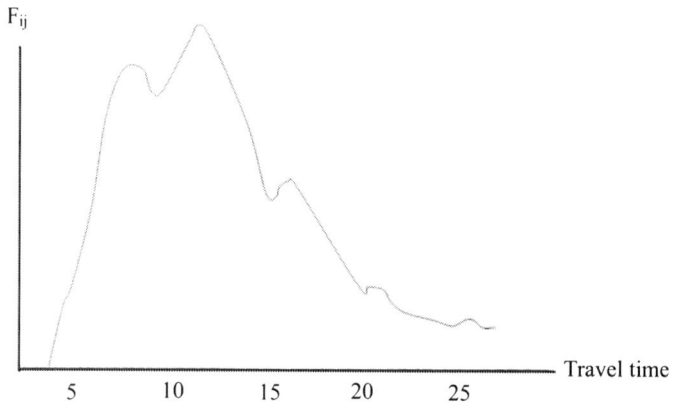

Figure 6.5. Friction factor versus travel time.

and productions as the original trip table. Table 6.7 presents row and column factoring for the 3-TAZ example.

As the factors are almost equal to 1 and the total A_js and P_is are the same as observed, we stop the factoring procedure.

6.3.2.3 Calculating Zone-to-Zone Adjustment Factors

The final step is to find zone-to-zone adjustment factors, which are calculated as follows.

$$K_{ij} = \frac{T_{ij} \text{ (observed)}}{T_{ij} \text{ (calculated)}} \qquad \text{(Equation 6.2)}$$

K_{ij} values are calculated for the preceding example in Table 6.8.

The estimated F_{ij} and K_{ij} can be used for the design year to forecast trip interchange (T_{ij}) using Equation 5.3.

6.3.3 MODE CHOICE CALIBRATION

The calibration process in the mode choice step is to estimate coefficients of utility functions for each mode in Equation 4.8. The calibration is performed using household travel survey data or stated and revealed preferences survey questionnaires specifically prepared for modal choice behavior. Other data to be utilized is transit boarding counts and census data. Other important data to obtain is daily person trips by mode stratified by trip purpose. Although many variables may affect mode choice, most of the current models utilize two major variables, time and cost. Both out-of-vehicle time and in-vehicle time must be considered in the model because studies show that travelers have different values for these two times and prefer to be in the vehicle, rather than waiting outside.

A logistic regression model is applied to the collected data (probability of choosing each mode) for the study area to find the variables affecting travelers' mode choice behavior and estimate the coefficients of the variables.

A reasonableness check would be to compare the estimated coefficients and modal shares to national or other regional ones. The reasonableness of the predicted mode choice probabilities for subgroups of the population, such as households without any vehicles and low-income households, must also be checked. Bus route assigned volumes need to

Table 6.7. Row and column factoring for the 3-TAZ example

| | Column factoring | | | |
	Trip table for base year			
F/T	1	2	3	p_i
1	39	93	167	300
2	45	12	43	100
3	86	46	18	150
A_j	171	151	228	550
Observed	200	160	190	
Factor	= 200/171 = 1.173	= 160/151 = 1.058	= 190/228 = 0.832	

| | | | Row factoring | | | |
			Trip table for base year			
F/T	1	2	3	p_i	Observed	Factor
1	46	99	139	284	300	= 300/284 = 1.056
2	53	13	36	101	100	= 100/101 = 0.987
3	101	48	15	164	150	= 150/164 = 0.912
A_j	200	160	190	550		

| | Column factoring | | | |
	Trip table for base year			
F/T	1	2	3	p_i
1	49	104	147	300
2	52	13	35	100
3	92	44	14	150
A_j	193	161	196	550
Observed	200	160	190	
Factor	1.036	0.993	0.970	

| | | | Row factoring | | | |
			Trip table for base year			
F/T	1	2	3	p_i	Observed	Factor
1	51	104	142	297	300	1.011
2	54	13	34	101	100	0.993
3	95	44	13	153	150	0.983
A_j	200	160	190	550		

Table 6.8. Zone-to-zone factoring (K) calculation for the 3-TAZ example

F/T	1	2	3
1	40/51 = 0.782	110/105 = 1.050	150/142 = 1.041
2	50/54 = 0.932	20/13 = 1.595	30/34 = 0.887
3	110/95 = 1.174	30/44 = 0.697	10/13 = 0.754

be compared to actual ridership. Furthermore, average auto occupancies by trip purpose, home-based work (HBW) transit trips as a percent of total transit trips, mode share to or from area types or major districts, and average auto occupancies to or from area types or major districts must be checked. A rule of thumb for reasonableness of times is that the ratio of the out-of-vehicle travel time coefficient to the in-vehicle travel time coefficient should be between 2.0 and 3.0. Also, time spent in a car for accessing transit is perceived as 1.5 times as burdensome as time spent in the transit vehicle itself. Similarly, time spent walking to access transit is perceived as twice as burdensome as time spent in the transit vehicle. Furthermore, a 10 percent transit fare increase will decrease ridership by 3 percent. These rules of thumb are used by modelers to check or calibrate the coefficients of the model.

6.3.4 TRAFFIC ASSIGNMENT CALIBRATION

The calibration in the traffic assignment step is composed of two different components. The first component is estimating the parameters of the link performance function using collected travel time data from different road types in the study area if the modeler chooses to use different parameters than the original Bureau of Public Roads (BPR). The second component is calibrating or tweaking model assumptions to reproduce counted traffic volumes for the base year. The major sources of traffic volume data are traffic counts and highway performance monitoring system (HPMS) data. HPMS is a national-level highway information system that contains information about the extent, condition, performance, usage, and operating characteristics of highways.

Traffic volume calibration must be performed for as many roadway segments of the network as possible. The rule of thumb is one-third of the roadway segments that are to be included in the coded network. Besides individual road segments, some aggregate checks must be performed as follows.

6.3.4.1 Vehicle Miles Traveled (VMT) Check

The vehicle miles traveled (VMT) produced by the model should be compared with observed VMT from traffic count and HPMS data. When traffic counts are being used, one must compare the estimated VMT for only those links in the network for which a count is available. However, when HPMS is used, comparison must be done for all links in the network. The VMT checks must be performed for the whole region as well as by market segments such as facility type, area type, and geographic subdivision. The difference between estimated and observed VMT should not exceed 5 percent.

6.3.4.2 Screen Lines, Cut Lines, and Cordon Counts

A model might not present exactly each link's volume; however, it must present the total volume on each screen line, cut line, or cordon.

Screen lines extend across the whole study area from one boundary line to the other boundary line. Screen lines are usually natural or manmade barriers such as a river or railroad, or jurisdictional boundaries. Cut lines extend across a corridor and do not cover the whole study necessarily. A screen line is a cut line, but a cut line is not necessarily a screen line. A cordon line covers the whole designated area such as a central business district (CBD). Figure 6.6 presents a screen line, cut line, and cordon line.

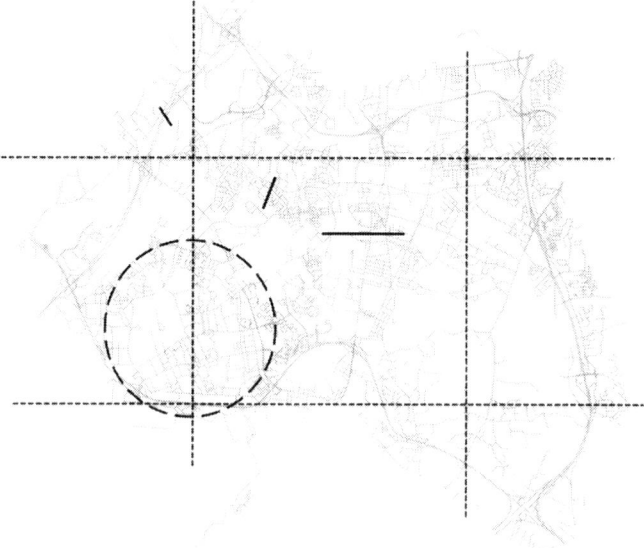

Figure 6.6. An example of screen line (--), cut line (___), and cordon (circle).

Screen line or cut line evaluations are performed by comparing the total counted screen lines or cut lines with the total estimated ones. A study area may have several screen lines (usually four to six) and tens of cut lines. The estimated screen lines must be within 5 percent of the counted ones, while the estimated cut lines must be within 10 percent.

6.3.4.3 Calibration Measurements

The following statistics are usually used to find whether the model is calibrated. The formulas can be used for individual links as well as groups of links.

$$r = \frac{\sum x \cdot y - n \bar{x} \bar{y}}{\sqrt{\left(\sum x^2 - n\bar{x}^2\right)\left(\sum y^2 - n\bar{y}^2\right)}} \qquad \text{(Equation 6.3)}$$

$$RMSE = \frac{\sqrt{\dfrac{\sum (x-y)^2}{n}}}{\dfrac{\sum x}{n}} \times 100\%$$

$$AE = \frac{\sum |y-x|}{\sum x} \times 100\%$$

$$PE = \frac{\sum (y-x)}{n} \times 100\%$$

where,
r = correlation coefficient
$RMSE$ = root mean square error
AE = absolute error
PE = percent error region-wide
x = ground count
y = link volume
n = number of links

The Federal Highway Administration suggests guidelines for calibration (Ismart 1990), which are summarized in Table 6.9. The rule of thumb for RMSE values is to be less than 30 percent.

Besides the automobile counts, transit ridership counts by route are used for transit assignment calibration.

Table 6.9. FHWA guidelines for calibration

	FHWA guideline
Correlation coefficient	0.88
Percent error region-wide	5 Percent
Sum of differences by functional class	
Freeway	7 Percent
Principal arterial	10 Percent
Minor arterial	15 Percent
Collector	25 Percent

It should be noted that a successful calibration of the traffic assignment step is based on the successful calibration of each of the previous steps. If the previous steps are not calibrated correctly, the calibration of trip assignment will be meaningless.

6.4   AN EXAMPLE IN CALIBRATION AND VALIDATION

An example details the procedures explained in this book. The study area selected is Fort Meade, a 40-square-mile area located southwest of Baltimore in Anne Arundel County, MD. It includes the Fort George G. Meade Army base, Arundel Mills mall, Maryland Live! Casino, and Odenton Town Center that attract many trips. It is a small area to develop a precise travel demand model for; thus, a high percentage of trips are externals. Modeling a large region is time-consuming and complicated and out of the scope of this book. Fort Meade is part of the Baltimore Metropolitan Council (BMC) regional highway network, which includes 2,928 TAZs. As presented in Figure 6.7, there are 28 TAZs and 263 links in the study area. Of the 28 TAZs, 13 are external (i.e., outside the study area). All of the trips outside the study area are assumed to traverse one of these TAZs to enter the study area. The model was originally developed as part of a project funded by Maryland State Highway Administration (SHA). The model presented in this book is just for demonstration purposes and might not be precise enough for policy and planning usages. The base model is 2005 and the design years are 2010, 2020, and 2030. Both the four-step model and activity-based model are developed and explained next.

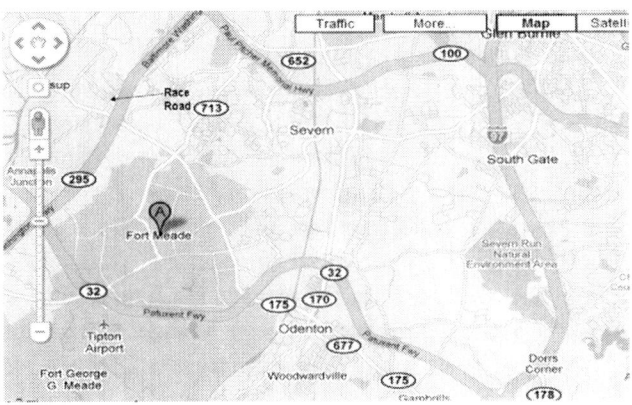

Figure 6.7. The study area.

6.4.1 FOUR-STEP MODEL CALIBRATION EXAMPLE

Because the study area is not an urban area and the dominant mode of transportation is auto without any significant transit, walk, or bike, it is assumed that the only available mode is car, and therefore, there is no mode choice step. Three models of AM peak, PM peak, and average annual daily traffic (AADT) were developed and the time-of-day classification was performed in the traffic assignment step.

A feedback loop was applied to the model. The traffic assignment module's outputs became the inputs for the trip distribution module, and a new pairing of origin and destination zones was performed for each trip type. This iteration was performed three times in this example to avoid oscillation. The first iteration used free-flow travel time for each link, assuming that there were no vehicles on the links; other iterations used the congested travel time obtained from the traffic assignment module. Therefore, the origin–destination pairing in trip distribution was more realistic using the congested travel time as impedance.

6.4.1.1 Data Preparation Example

Most of the data such as zoning and land use data was retrieved from the BMC model. The road network was developed in TransCAD software. Road information, such as speed limit and number of lanes, was coded in the software. Household travel survey 1994 was employed by the BMC model to estimate trip rates, to find trip length distribution, and to calculate trip distribution between TAZs.

6.4.1.2 Trip Generation Calibration Example

The trip generation module estimates person trips originating from and destined for each TAZ in the model. The input data includes land use data—which contains population, employment, retail employment, non-retail employment, and number of households for each TAZ—obtained from the 2005 socio-economic data from the BMC model. The regional BMC model developed a cross-classification model for productions using income, household size, or number of workers in the household (depending on the trip purpose), and a linear regression model for trip attractions to find trip rates. Although BMC used seven trip purposes, we used only three trip purposes because our study area is much smaller and there are many external trips. We also used TransCAD's default trip rates to estimate trip productions and attractions using the land use file obtained from the BMC. Table 6.10 shows the land use file, including population, number of household, median income per household, and number of workers in each employment segment (retail, office, industry, and other) for each TAZ in the study area. The first 15 TAZs are internal and the last 13 TAZs are external, for which no information is available. The number of productions and attractions is assumed for each external TAZ to produce external trips on the major roads entering the area. Trip generation finds the number of trips produced from and attracted to each of the internal TAZs using the land use file and trip rates for each trip purpose. Then, trip attractions were balanced to trip productions as explained in Chapter 5. Table 6.11 presents part of the estimated productions and attractions for the study areas by trip purpose. External TAZs' production and attraction for HBW trips comes directly from the land use file assuming all external trips are work trips. The output is compared to the regional BMC model outputs to make sure the results are reasonable.

6.4.1.3 Trip Distribution Calibration Example

The trip distribution module uses a gravity model with K-factors (Equation 4.3) to predict the number of trips between each TAZ pair for each trip purpose.

The trip distribution module's input is the production–attraction (PA) table, an impedance matrix, and trip table and trip length distribution for the base year obtained and calculated from the household travel survey. The PA table is the trip generation's output (as explained earlier). The impedance matrix includes auto travel times between zones, intrazonal times, and terminal times. Travel time is free-flow travel time in the first iteration and

Table 6.10. Land use file for the study area

TAZ	Population	Household	Med-inc	Workers	Retail	Office	Industry	Other	Non-retail	Acres	External	Pro-duction	Attrac-tion
1	2,484	854	84,463	1,304	4	62	19	26	107	404	0		
2	2,237	801	76,164	1,271	9	107	23	71	201	616	0		
3	1,399	534	65,484	747	3	21	5	9	35	686	0		
4	6,515	73	59,043	172	361	3,566	429	1,199	5,194	2,248	0		
5	1,981	684	77,759	954	76	99	6	58	163	195	0		
6	2,889	925	81,758	1,699	13	68	9	22	99	590	0		
7	4,674	1,767	54,082	2,342	9	86	13	83	182	444	0		
8	4,072	1,505	66,167	2,081	20	1,198	699	394	2,291	1,137	0		
9	3,398	1,130	32,399	1,356	7	38	7	9	54	433	0		
10	9,144	3,230	69,080	4,420	365	148	40	59	247	835	0		
11	150	62	50,926	94	20	332	124	217	673	303	0		
12	3,444	1,364	59,562	2,115	163	1,309	397	417	2,123	1,139	0		
13	340	135	38,612	213	377	217	28	195	440	141	0		
14	9,936	2,500	42,928	3,301	348	17,419	1,145	29,338	47,902	13,902	0		
15	297	80	51,330	111	13	538	83	663	1,284	1,479	0		
87											1	35,000	18,000

(Continued)

Table 6.10. (Continued)

TAZ	Population	Household	Med-inc	Workers	Retail	Office	Industry	Other	Non-retail	Acres	External	Pro-duction	Attrac-tion
88											1	18,000	22,000
89											1	162,000	165,000
90											1	25,600	18,600
91											1	22,000	52,000
92											1	48,000	32,000
93											1	10,000	12,000
94											1	80,000	90,000
95											1	9,000	18,000
96											1	4,500	4,500
97											1	230,000	180,000
98											1	25,000	40,000
99											1	170,000	166,000

Table 6.11. Part of the estimated productions and attractions by trip purposes for the base model

TAZ	HBW_P	HBW_A	HBNW_P	HBNW_A	NHB_P	NHB_A
1	2476.6	117.3	7058.3	1913.6	2848.1	223.6
2	2322.9	222.0	6620.3	2002.4	2671.3	293.2
3	1548.6	40.2	4413.5	1174.4	1780.9	114.8
4	211.7	5872.2	603.3	12683.1	243.5	4375.2
5	1983.6	252.6	5653.3	3080.9	2281.1	287.0
6	2682.5	118.4	7645.1	2230.6	3084.9	234.3
7	5124.3	201.9	14604.3	3934.2	5892.9	431.8
8	4364.5	2442.9	12438.8	5756.9	5019.2	2075.6
9	3277.0	64.5	9339.5	2478.0	3768.6	227.3
10	9367.0	646.9	26696.0	14144.3	10772.1	943.1
...						
99	170000.0	166000.0	0	0	0	0

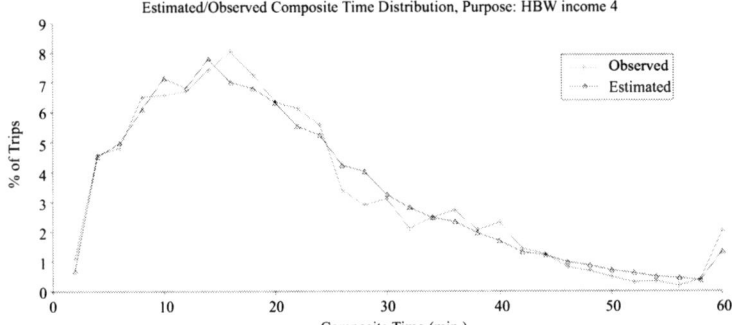

Figure 6.8. A trip length distribution sample from the BMC (2005) model.

in other iterations is estimated by traffic assignment. Intrazonal time is a direct input per zone, and terminal time is the time assumed between leaving the location and starting the trip. Terminal time was assumed to be two min for internal trips and 10 minutes for external trips, based on the size of the area. Trip length distribution from the BMC model for each trip purpose (Figure 6.8) was adopted in this model to estimate model parameters. The trip table for the base model was obtained and adjusted from the 1994 household travel survey to find the friction factors for the base year. The following formula was used to create a friction factor matrix for each trip purpose. Modelers may use a different formula for F_{ij} or even use tables,

Table 6.12. Part of the estimated trip table for the HBW of the base model

	1	2	3	4	5	6	7	8	9	10
1	35.6	16.2	0.6	28.5	4.8	3.5	9.0	47.4	1.9	5.9
2	7.0	55.2	1.0	37.4	8.3	3.3	5.0	19.5	0.7	4.7
3	0.7	2.7	4.5	39.9	6.4	1.5	1.5	7.1	0.4	3.5
4	0.0	0.1	0.0	41.4	0.1	0.1	0.1	0.4	0.0	0.2
5	1.2	4.9	1.4	34.5	34.0	3.3	2.8	11.6	0.6	5.3
6	3.0	6.6	1.1	43.5	11.3	25.5	7.2	25.2	1.0	6.3
7	11.3	14.6	1.5	71.0	13.8	10.6	82.0	75.9	3.0	10.0
8	2.6	2.5	0.3	30.5	2.5	1.6	3.3	401.0	1.2	7.2
9	3.2	2.9	0.6	42.1	4.1	1.9	3.9	36.6	11.3	15.2
10	3.1	5.7	1.5	103.2	11.0	3.9	4.2	59.8	4.8	181.9

rather than a formula. The coefficients are different for each trip purpose, and they were estimated using the base year trip table.

$$f_{ij} = e^{(C_1 + C_2 . d_{ij} + C_3 . \ln(d_{ij}))}$$ (Equation 6.4)

where,

f_{ij} = the friction factor between zone i and zone j
d_{ij} = the impedance between zone i and zone j
C_{1-3} = constants calibrated for each trip type

The preceding procedure calibrates the trip distribution model (Equation 4.3), as explained in Section 6.2 to represent trip tables for the base year. The module's output was zone-to-zone person trips by trip purpose. The trip distribution's output was three matrixes of the number of trips between zones for each trip purpose. Table 6.12 presents part of the trip table.

6.4.1.4 Traffic Assignment Calibration Example

As stated earlier, there is no mode choice step, and trip distribution output is the input to traffic assignment. The traffic assignment module estimates traffic flow on each of the network's roads (links). The module's input is a flow matrix that indicates the traffic volume between each origin and destination (OD) zone. The OD flow matrix is calculated by converting PA to OD (explained in Chapter 6), person trips to vehicle trips for each trip purpose, converting daily trips to peak hour trips, and aggregating zone-to-zone person trips by trip purpose (the three trip distribution output matrixes). In other words, traffic assignment makes three OD matrixes: ADT, AM peak, and PM peak hour. We converted person trips to vehicle trips by assuming average auto occupancy of 1.08 based on year 2000 census data for the state of Maryland. To calculate the AM and PM peak hour, we multiplied the AM peak period (6:00 to 9:00) and PM peak period (16:00 to 19:00) by 0.4 and 0.38, respectively, based on the BMC model.

We utilized a user equilibrium model to assign the OD flow based on each route's travel cost. The travel cost is a combination of travel time and distance, and each route is a set of links that connects the origin and destination. A BPR volume delay function (Equation 6.9) is utilized. The default values for α, β parameters are 0.15 and 4, respectively, for all road functional classes. The traffic assignment module's outputs are the link volumes, link travel times, volume over capacity (VOC) ratio, VMT, vehicle hours travelled (VHT), and link speeds for each direction (AB and BA). Table 6.13 presents part of the output.

Ground counts were obtained for approximately 50 percent of the links in the study area (158 counts out of 327 links). Individual link errors were calculated by subtracting the estimated model volume from the link's ground count. Several iterations of calibration were performed until the results are within FHWA guidelines. The items to be revised are road characteristics in the network by functional class, such as BPR parameters (α, β), capacity of the road, and FFS (it was originally assumed to be 20 percent higher than the speed limit). We also checked zero volume links and highly congested links (V/C > 1) and fixed the problems if they did not reflect the real world. The major links entering or exiting the study area are those on which external trips are carried and should have almost the exact link volumes.

This is fixed by revising the land use file (Table 6.10) for external zones. If trips are very high or very low in the whole study area, checking and revisions in trip generation and trip distribution need to be performed. Tables 6.14 to 6.16 compare the FHWA guidelines and the calibrated AM, PM, and ADT models. VMT and VHT are also reported for each model, which must be checked against the observed data.

Table 6.13. Part of traffic assignment output

Link-ID	AB_Flow	BA_Flow	AB_Time	BA_Time	AB_VOC	BA_VOC	AB_VMT	BA_VMT	AB_VHT	BA_VHT	AB_Speed	BA_Speed
597	55.61	22.97	0.82	0.82	0.00	0.00	18.98	7.84	45.54	18.81	25.00	25.00
598	300.51	129.78	1.15	1.15	0.01	0.01	143.45	61.95	344.29	148.69	25.00	25.00
599	10.61	9.72	1.08	1.08	0.00	0.00	4.78	4.37	11.46	10.50	25.00	25.00
600	366.35	138.86	0.85	0.85	0.02	0.01	130.16	49.34	312.39	118.41	25.00	25.00
603	320.26	135.14	0.78	0.78	0.02	0.01	104.43	44.07	250.63	105.76	25.00	25.00
604	98.73	35.65	0.83	0.83	0.00	0.00	34.02	12.29	81.66	29.48	25.00	25.00
605	762.90	286.41	0.91	0.91	0.04	0.01	287.92	108.09	691.00	259.42	25.00	25.00
606	35.00	31.48	1.01	1.01	0.00	0.00	14.75	13.26	35.40	31.83	25.00	25.00
607	54.50	14.49	0.69	0.69	0.00	0.00	15.69	4.17	37.66	10.02	25.00	25.00
608	234.62	110.60	1.22	1.22	0.01	0.01	118.97	56.09	285.54	134.60	25.00	25.00
609	513.82	202.85	1.01	1.01	0.02	0.01	215.35	85.02	516.84	204.04	25.00	25.00
612	5.11	446.40	1.66	1.66	0.00	0.02	4.23	369.75	8.46	739.49	30.00	30.00
613	259.65	366.34	2.08	2.08	0.01	0.02	269.72	380.55	539.43	761.09	30.00	30.00

616	140.24	73.32	1.08	1.08	0.01	0.00	63.21	33.04	151.70	79.31	25.00	25.00
617	106.41	27.31	1.50	1.50	0.01	0.00	66.67	17.11	160.00	41.07	25.00	25.00
630	581.69	395.69	0.67	0.67	0.03	0.02	161.93	110.15	388.62	264.35	25.00	25.00
697	55.87	166.14	0.94	0.94	0.00	0.01	26.13	77.71	52.27	155.42	30.00	30.00
698	281.22	2206.85	1.28	1.28	0.01	0.11	180.26	1414.59	360.52	2829.23	30.00	30.00
699	152.15	1730.86	1.10	1.10	0.01	0.08	83.56	950.64	167.13	1901.29	30.00	30.00
701	499.88	350.91	0.93	0.93	0.02	0.02	193.18	135.61	463.64	325.47	25.00	25.00
702	151.95	194.68	0.25	0.25	0.01	0.01	16.08	20.60	38.60	49.45	25.00	25.00
703	8.30	11.28	0.25	0.25	0.00	0.00	0.86	1.17	2.08	2.82	25.00	25.00
705	73.87	35.58	0.62	0.62	0.00	0.00	19.20	9.25	46.09	22.20	25.00	25.00
706	1277.40	471.59	0.98	0.98	0.06	0.02	519.83	191.91	1247.59	460.58	25.00	25.00

Table 6.14. AM peak hour calibration for the study area

	FHWA guideline	Model
Correlation coefficient	0.88	0.95
Percent error region-wide	5%	−4.8%
Sum of differences by functional class		
Freeway	7%	−2.8%
Principal arterial	10%	−6.7%
Minor arterial	15%	−13.47%
Collector	25%	

VMT: 189,660.7
VHT: 4,706

Table 6.15. PM peak hour calibration for the study area

	FHWA guideline	Model
Correlation coefficient	0.88	0.92
Percent error region-wide	5%	−2.3%
Sum of differences by functional class		
Freeway	7%	−3.6%
Principal arterial	10%	2.9%
Minor arterial	15%	−9.8%
Collector	25%	

VMT: 198,279
VHT: 4,965.9

Table 6.16. ADT calibration for the study area

	FHWA guideline	Model
Correlation coefficient	0.88	0.93
Percent error region-wide	5%	−4.4%
Sum of differences by functional class		
Freeway	7%	−5.5%
Principal arterial	10%	−11.4%
Minor arterial	15%	12.7%
Collector	25%	

VMT: 2,067,222.9
VHT: 46,159.1

6.4.2 ACTIVITY-BASED AND AGENT-BASED MODEL CALIBRATION EXAMPLE

An activity-based and agent-based model for the same study area was created in TRANSIMS. The model is based on travelers' activities and includes all local roads and driveways, rather than only major roads. The household activity survey was adopted from the national household activity survey. The land use data was acquired from the four-step model.

TRANSIMS is based on individual behavior and interactions. It traces and simulates the movements of each individual in a fully described network as he or she accomplishes tasks in a 24-hour period. TRANSIMS also collects statistics on traffic, congestion, and pollution. TRANSIMS is more data-intensive than TransCAD. Figure 6.9 presents a schematic view of TRANSIMS. We developed two different models in TRANSIMS: Track-1 and Track-2.

Track-1 includes only route planner, microsimulator, and emission estimator and receives OD matrixes from a four-step model, rather than generating synthetic households and activities. Therefore, Track-1 is not activity-based, rather it is trip-based. TRANSIMS' Track-1 has been used by several researchers and practitioners. Although this model is not activity-based, it can be considered a dynamic four-step model. The trip-based model breaks the OD trip matrixes into different times of day and uses the time-dependent shortest path (route planner) to load the trips on the network. It also simulates network traffic. As the trip-based model is dynamic and uses a microsimulator, it addresses the problems in the four-step model's assignment step.

The Track-2 model, which is an activity-based model, uses all modules of TRANSIMS and the national household activity survey. The Track-2 model may underestimate traffic volumes because it does not convert external trips to activities, and therefore, does not include external trips. This issue does not affect models for large areas because external trips are negligible compared to the areas' trips. However, for such a small study area, one where external trips are non-negligible, this is problematic.

Figure 6.10 presents the links, nodes, and zones of the TRANSIMS-converted network. The network is more detailed than in the four-step model. It contains 1,782 links and 1,461 nodes. Figure 6.11 presents the activity locations on the TRANSIMS network.

6.4.2.1 Population Synthesizer

The first TRANSIMS module, population synthesizer, generates the entire population in the study area using census (PUMS, PUMA, and STF-3A),

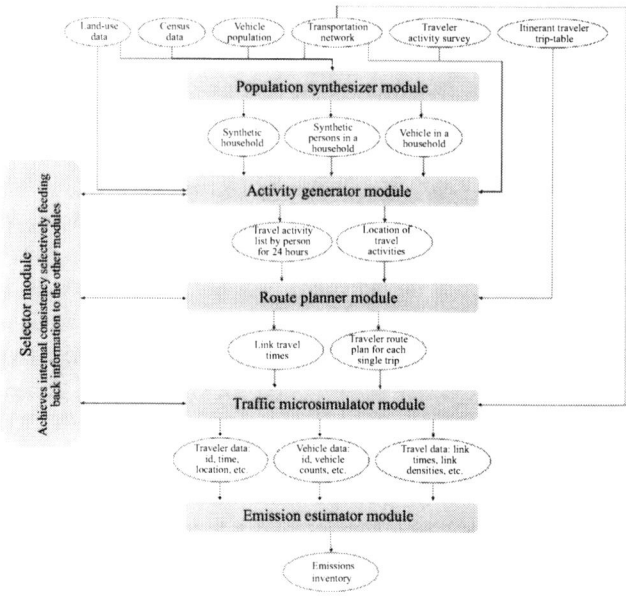

Figure 6.9. TRANSIMS framework (Hobeika 2010).

Figure 6.10. Links, nodes, and zones for the TRANSIMS-converted network for the study area.

survey, land use, and network data. PUMS and PUMA were explained in Chapter 5. STF-3A contains a demographic summary table for small geographic areas, census tracts, or census block groups.

TRANSIMS uses an algorithm developed by Beckman et al. (1996) to generate a synthetic population with characteristics very similar to the households found in a household survey. It uses a two-step iterative proportional fitting (IPF) procedure to create a cross-classification of

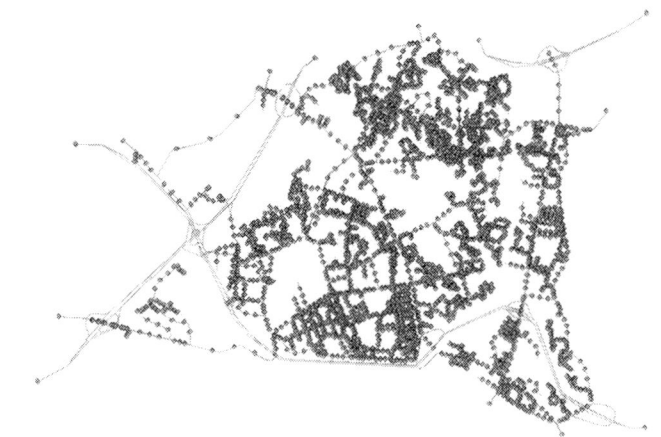

Figure 6.11. Activity locations for the study area in TRANSIMS.

household characteristics that matches the distribution of key household attributes specified for a location.

For each synthetic household, TRANSIMS specifies the household location and socioeconomic characteristics such as size, number of vehicles, and income, as well as demographic characteristics of each household member such as age and gender as presented in Tables 6.17 and 6.18. Land-use data are used to locate each household at an activity location on the transportation network.

6.4.2.2 Activity Generator

The second module is the activity generator that uses activity surveys to generate a list of activities for a 24-hour period for each member of the synthetic households. Each activity consists of an activity type and its priority, starting and ending time preferences, a preferred mode of transportation, a vehicle preference (if applicable), a list of possible locations, and a list of other participants (if shared). It uses location choice models to specify activity location for non-home activities. Unlike four-step models that utilized trip types such as HBW, HBO, and NHB, an activity-based model includes different activities such as home, work, school, visit, shop, and so on. Fourteen different travel modes (walk, bike, car, bus, rail, and so on) were embedded in TRANSIMS.

The most important input to the activity generator is the household activity survey data that includes detailed activity information for each

Table 6.17. Sample household file

HHOLD	Location	Persons	State	PUMA	Vehicles	Workers	Income	HHAGE	NUM_LT5	NUM_5TO15
1	420	3	24	1201	2	2	3,500	19	1	0
2	2393	3	24	1201	2	2	3,500	19	1	0
3	2340	3	24	1201	2	2	3,500	19	1	0
4	559	3	24	1201	1	1	1,600	29	0	2
5	2381	3	24	1201	1	1	1,600	29	0	2

Table 6.18. Sample person file

HHOLD	Person	Age	Gender	Worker
1	1	19	1	1
1	2	19	2	1
1	3	1	1	0
2	1	19	1	1
2	2	19	2	1

member of the household for a 24-hour period. This survey is very similar to the travel survey explained in Chapter 5. Table 6.19 presents a sample household activity survey.

The activity generator utilizes the Classification and Regression Tree (CART) algorithm introduced by Breiman et al. (1984) to group survey households based on their demographic information. It produces a classification of household demographic characteristics (such as household size, household income, and so on) based on households' travel behaviors (such as time spent at home, work, shopping centers, and so on). The CART algorithm makes a binary matching tree and locates households in the terminal node of the tree, which is the end of the path of the selected household characteristics. The tree is sensitive to the characteristics of household behavior. Figure 6.12 shows a simple sample of a binary tree. The tree includes 13 nodes, seven of which are terminal nodes shown by squares and the remaining six nodes are non-terminals shown by circles. At each non-terminal node, a household is classified further into either a left or right node according to the demographic variable.

After developing a classification tree, the activity generator module matches the given synthetic household with a survey household, generates activity times and durations, and then creates activity locations. It uses a destination choice model to locate non-home activities. Figure 6.13 presents an example of the location choice in the activity generator. Table 6.20 presents part of the activity generator output for the study area.

6.4.2.3 Route Planner

The route planner creates travel plans for each individual person in the network using the activity generator output. Unlike four-step models, TRANSIMS models trip chains. Each trip chain includes several travel legs, each of which has travelers' ID, start and destination location, start

Table 6.19. Sample household activity survey data

HHOLD	PERSON	ACTIVITY	PURPOSE	PRIORITY	START	END	DURATION	MODE	VEHICLE	LOCATION	PASSENGER	CONSTRAINT
200007	1	1	0	0	0:00	10:33	10:33	1	0	1	0	0
200007	1	2	1	1	11:48	14:30	2:42	2	2	2	0	1
200007	1	3	1	2	14:45	19:00	4:15	2	2	3	0	0
200007	1	4	5	0	19:45	22:45	3:00	2	2	4	0	1
200007	1	5	0	0	23:30	1@3:00	3:30	2	2	5	0	5
200009	1	6	0	0	0:00	10:00	10:00	1	2	1	1	2
200009	1	1	12	2	10:10	12:10	2:00	2	0	2	0	0
200009	1	2	12	0	12:15	13:23	1:08	2	1	3	1	2
200009	1	4	15	0	13:45	1@3:00	13:15	2	1	4	0	1
200010	1	4	0	0	0:00	10:39	10:39	1	0	1	3	0
200010	1	5	13	0	11:10	11:25	0:15	8	0	2	3	1
200010	1	6	12	2	11:50	13:20	1:30	8	1	3	0	0
200010	1	1	15	0	13:40	14:30	0:50	2	0	4	0	0
200010	1	2	12	0	14:45	15:00	0:15	2	4	5	0	0
200010	1	3	0	0	15:21	23:30	8:09	2	4	6	0	0
200010	2	4	9	2	0:00	14:25	14:25	1	4	1	2	2
200010	2	5	0	0	14:45	15:35	0:50	2	4	2	2	2
200010	2	6	0	0	16:00	16:55	0:55	2	4	3	2	2
200010	2	7	13	0	17:10	17:11	0:01	2	4	4	2	2
200010	2	8	12	2	17:30	20:00	2:30	2	4	5	1	2

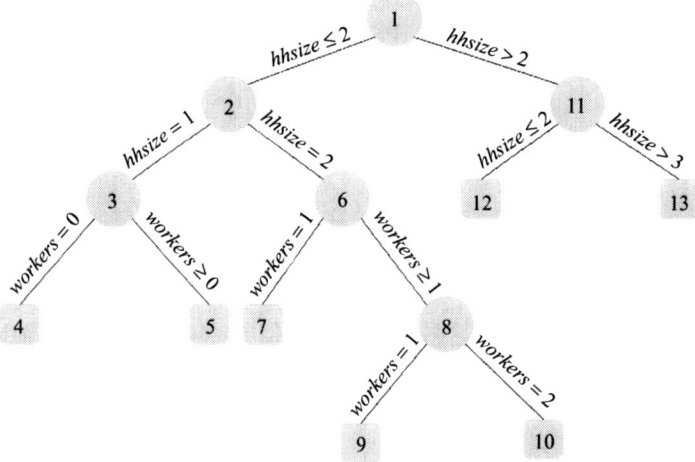

Figure 6.12. A simple binary tree.

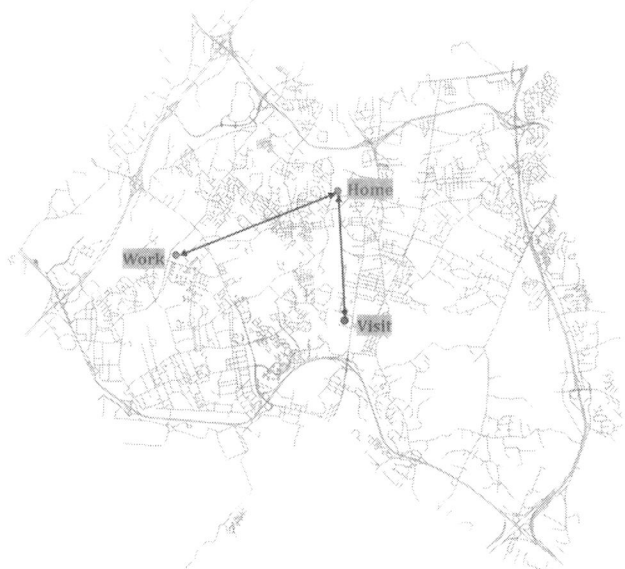

Figure 6.13. Example of the activity generator's location choice.

time, maximum travel time, and mode string. The route planner utilizes a modified version of Dijkstra's algorithm (Dijkstra 1959), which is a time-dependent, label-constrained shortest path (TDLSP). Figure 6.14 presents a route planner shortest path example. While the original Dijkstra's algorithm is not time-dependent, departure time is taken into account in

Table 6.20. Activity generator output sample

HHOLD	PERSON	ACTIVITY	PURPOSE	PRIORITY	START	END	DURATION	MODE	VEHICLE	LOCATION	PASSENGER	CONSTRAINT
1	1	1	0	0	0:00	7:13:07	7:13:07	1	0	418	0	0
1	1	2	9	2	7:30	11:45	4:15	2	2	3964	0	1
1	1	3	0	0	12:01:53	12:45	0:43:07	2	2	418	0	0
1	1	4	9	2	13:00	16:00	3:00	2	2	3159	0	1
1	1	5	6	0	16:15	16:20	0:05	2	2	3904	0	5
1	1	6	0	0	16:30	1:03:00	10:30	1	0	418	1	2
1	2	1	0	0	0:00	6:57:10	6:57:10	2	1	418	0	0
1	2	2	6	0	7:15	7:20	0:05	2	1	3904	1	2
1	2	3	9	2	8:00	12:00	4:00	2	1	3520	0	1
1	2	4	5	0	12:05:12	12:45	0:39:48	8	0	3842	3	0
1	2	5	9	1	13:00	17:00	4:00	8	0	3831	3	1
1	2	6	0	0	17:33:55	1:03:00	9:26:05	8	1	418	0	0
1	3	1	0	0	0:00	6:57:10	6:57:10	1	0	418	0	0
1	3	2	7	2	7:15	16:20	9:05	10	1	3904	1	2
1	3	3	0	0	16:30	1:03:00	10:30	10	2	418	1	2
2	1	1	0	0	0:00	9:44:27	9:44:27	1	0	2413	0	0
2	1	2	14	2	10:05	13:00	2:55	2	4	3378	0	0
2	1	3	0	0	13:20	13:48:20	0:28:20	2	4	2413	0	0
2	1	4	2	0	14:05	14:25	0:20	2	4	3043	2	2
2	1	5	3	2	14:45	15:35	0:50	2	4	3426	2	2
2	1	6	0	0	16:00	16:55	0:55	2	4	2413	2	2
2	1	7	16	2	17:10	17:11	0:01	2	4	3318	2	2
2	1	8	0	0	17:30	20:00	2:30	2	4	2413	1	2

the TDLSP. The label-constraint is related to the mode of travel. A collection of labors is a mode string such as WCW (walk–car–walk) and WBWRW (walk–bus–walk–rail–walk).

The route planner transforms the network to a set of interconnected, unimodal layers, each of which is associated with a mode (Figure 6.15). Then, an imaginary link, which is called a process link, connects the unimodal layers to each other. Based on individual traveler preferences and constraints specified in the activity generator output, the route planner specifies all possible sequences of the travel modes (mode strings) by which one can reach the destination from the starting node. Then, an internal network that consists of links and nodes on all the layers is made. Each link in the internal network has an associated travel time.

Initially, it is assumed that there is no vehicle traffic in the network, and thus the FFS is utilized to calculate the travel times. The route planner then finds a shortest path for the trips in the internal network based on departure time and produces plans for all travelers in the network. The traffic microsimulator module uses the travelers' plans to simulate each individual and their interactions in the network. Then, travel time of each link is calculated to be used in the subsequent route planner runs in order to find the shortest path when the network includes all travelers. The travel time is calculated every 15 min. Jeihani, Sherali, and Hobeika (2006)

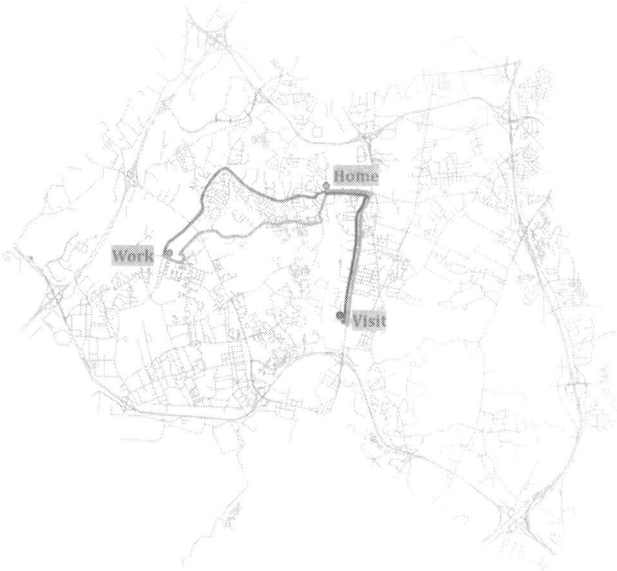

Figure 6.14. Example of the route planner shortest path.

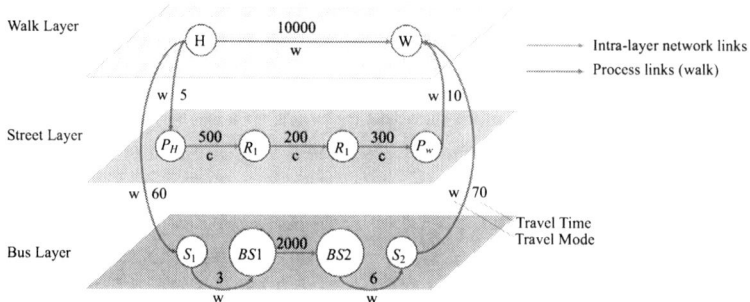

Figure 6.15. The internal network consisting of multimodal layers.

updated link travel time after each assignment, rather than assigning all travelers first.

The output of the route planner is a travel plan that includes the following information for each individual in the network:

- Traveler ID
- Trip ID
- Leg ID
- Starting time
- Starting location ID
- Starting accessory type
- Ending location ID
- Ending accessory type
- Duration
- Stop time
- Monetary cost
- Generalized cost function
- Max time flag
- Driver flag
- Travel mode (car, transit, pedestrian, and so on)
- Number of tokens
- Vehicle ID
- Number of passengers
- Token ID

6.4.2.4 Traffic Microsimulator and Visualization

The microsimulator module simulates second-by-second travelers' movements and interactions throughout the study area for the whole 24-hour period based on the individual travel plans provided by the route planner and computes the transportation system dynamics. The microsimulator is a powerful tool that is capable of handling intermodal travel plans, multiple travelers per vehicle, multiple trips per traveler, and vehicles with different operating characteristics. It is agent-based and capable of simulating detailed information on large networks using the cellular automation (CA) concept. It divides each link on the network into a finite number of cells that are 24.6 ft (7.5 m) long. Therefore, the speeds are measured in 24.6 ft per second (7.5 m per second) increments. The network is detailed and includes all streets, lanes, pocket lanes, transit stops, parking spots and garages, and so on. Vehicles can accelerate, decelerate, turn, change lanes, pass, respond to traffic controls, and interact with other vehicles in the simulator. Figure 6.16 presents a section of a network in the microsimulator.

The microsimulator gives vehicle snapshot data, traveler event, and summary data as output. The vehicle snapshot data includes the position of each vehicle on the network for every second. The event data yields detailed information for every event (Trip ID, Leg ID, Time, Location, and so on) for each individual in the network. The summary data includes spatial and temporal data over roadway segments. Some of the data in the summary file are flow, density, travel time, and so on. Figures 6.17 to 6.19 present samples of the microsimulator outputs.

The vehicle snapshot data can be presented using visualization software such as NEXTA (Zhu 2010). NEXTA reports minute-by-minute, link-specific vehicle spatial information, volumes, speed, travel time, and bottlenecks. Figure 6.20 presents a snapshot of the NEXTA output for the study area's PM peak. The red lines in this figure are the extremely congested areas in which the average speed in the corresponding streets is less than half of the speed limit.

6.4.2.5 Feedback Module and Calibration

The feedback process is the calibration tool in TRANSIMS. Feedback can be run between two or more modules. Feedback is used to calibrate the model, stabilize travel times in the network, yield the desired mode choice, and correct network, locations, modes, and activity times. Figure 6.21 presents different feedback loops.

Several feedbacks were performed for the activity generator to clean up the activity plans. Also, a number of feedbacks between the route planner and activity generator were run to correct the network and correct the discontinuity problems between the links and process links.

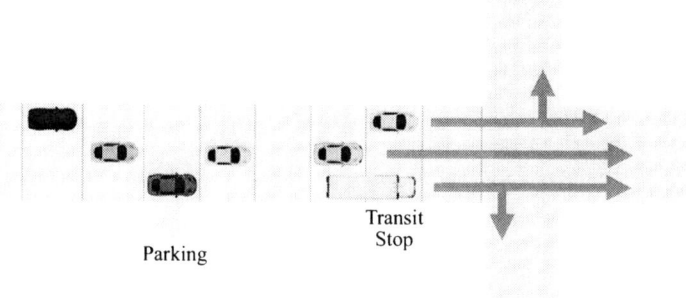

Parking

Transit
Stop

Figure 6.16. TRANSIMS' microsimulator.

The feedback between the route planner and the microsimulator was performed 11 times to stabilize travel times in the study area. At the first iteration, the route planner used free-flow travel times to find the shortest path. However, after all vehicles were loaded onto the network, the link travel times were higher than the free-flow travel time, and thus, some routes were no longer the shortest path. The feedback randomly re-routed 15 percent of the travelers in each iteration until the link travel times stabilized.

6.4.2.6 Emission Estimator

The emissions estimator is the last module in TRANSIMS. It translates vehicular traffic interactions into emissions and energy consumption. This module estimates tailpipe emissions for heavy-duty vehicles and tailpipe and evaporative emissions for light-duty vehicles. In spite of other emission models, it considers the effect of transient power change.

LINK	DIR	START_TIME	END_TIME	AVG_VOL	IN_VOL	OUT_VOL	AVG_SPEED	AVG_TIME	AVG_DELAY	AVG_DENSITY	MAX_DENSITY	TIME_RATIO	AVG_QUEUE	MAX_QUEUE	NUM_FAIL	VMT	VHT	NCONNECT	OUT_LINK	OUT_DIR	OUT_TURN	OUT_TIME
2	1	12:45	13:00	2	2	2	11.23	7.8	0	0.03	0.03	1	0	0	0	175.2	15.6	3	4	1	2	7.8
																			99	0	0	7.8
																			580	1	0	7.8
2	1	13:15	13:30	1	1	1	11.23	7.8	0	0.01	0.01	1	0	0	0	87.6	7.8	3	4	1	1	7.8
																			99	0	0	7.8
																			580	1	0	7.8
2	1	13:45	14:00	2	2	2	11.23	7.8	0	0.03	0.03	1	0	0	0	175.2	15.6	3	4	1	2	7.8
																			99	0	0	7.8
																			580	1	0	7.8
2	1	14:00	14:15	2	2	2	11.23	7.8	0	0.03	0.03	1	0	0	0	175.2	15.6	3	4	1	2	7.8
																			99	0	0	7.8
																			580	1	0	7.8
2	1	14:15	14:30	2	2	2	11.23	7.8	0	0.03	0.03	1	0	0	0	175.2	15.6	3	4	1	2	7.8
																			99	0	0	7.8
																			580	1	0	7.8

Figure 6.17. Trip performance result for the 10th iteration of microsimulation for the study area.

HOUSEHOLD	PERSON	TRIP	LEG	MODE	EVENT	SCHEDULE	ACTUAL	LINK	OFFSET
27781	1	1	2	0	1	0:06:09	0:04:46	−1325	59
120126	1	1	2	0	1	0:07:56	0:05:06	1442	84
44976	1	1	2	0	1	0:08:32	0:05:18	1705	72
91052	1	1	2	0	1	0:11:50	0:05:24	1587	140
71492	1	1	2	0	1	0:08:24	0:05:52	−971	37
38343	1	1	2	0	1	0:14:40	0:05:58	−1256	480
27781	1	1	3	2	0	0:06:09	0:04:49	1325	176
27781	1	1	3	2	1	0:06:39	0:05:19	1325	176
29608	1	1	2	0	1	0:12:58	0:06:12	−1022	73
126237	1	1	2	0	1	0:09:34	0:06:20	−1348	369
131378	1	1	2	0	1	0:08:16	0:06:41	1273	201
41716	1	1	2	0	1	0:08:43	0:06:54	1595	58
101184	1	1	2	0	1	0:11:24	0:07:15	80	47
119921	1	1	2	0	1	0:09:39	0:07:15	1091	35
72067	1	1	2	0	1	0:08:42	0:07:36	1480	93
39761	1	1	2	0	1	0:12:18	0:07:44	−1088	346
135552	1	1	2	0	1	0:09:08	0:07:47	884	112
120126	1	1	3	2	0	0:07:56	0:05:06	−1442	84
120126	1	1	3	2	1	0:08:26	0:05:36	−1442	84
87502	1	1	2	0	1	0:17:46	0:07:58	−742	314
28851	1	1	2	0	1	0:09:22	0:08:01	−1256	480
83834	1	1	2	0	1	0:10:27	0:08:01	153	54
113681	1	1	2	0	1	0:11:40	0:08:15	−1614	61

Figure 6.18. Microsimulation event result for the 10th iteration for the study area.

VEHICLE	TIME	LINK	DIR	LANE	OFFSET	SPEED	ACCEL	VEH_TYPE	DRIVER	PASSENGERS
43599	7:00	1798	1	1	982.5	30	0	1	1790902	0
47290	7:00	751	1	1	71.3	7.5	−22.5	1	1956801	0
47288	7:00	568	1	1	15	15	7.5	1	1956803	0
7345	7:00	1436	1	1	270	30	0	1	325301	0
7903	7:00	1441	0	1	45	30	7.5	1	347801	0
44729	7:00	1430	1	1	210	30	0	1	1850002	0
47333	7:00	242	0	3	96.4	7.5	7.5	1	1957801	0
47332	7:00	242	0	3	82.5	0	0	1	1957802	0
46710	7:00	242	1	2	75	15	−15	1	1937202	0
8489	7:00	1492	1	1	22.5	30	0	1	367601	0
8488	7:00	1487	1	1	60	30	0	1	367602	0
8487	7:00	1487	1	1	135	30	0	1	367603	0
7779	7:00	201	1	2	405	22.5	−15	1	343201	0
7778	7:00	242	0	3	75	0	0	1	343203	0
47038	7:00	201	0	2	367.5	37.5	0	1	1949202	0
47814	7:00	201	0	2	210	22.5	7.5	1	1978302	0
46342	7:00	1357	1	1	97.5	30	0	1	1921206	0
45579	7:00	1429	1	1	172.5	30	0	1	1889103	0
45578	7:00	201	0	2	247.5	30	−7.5	1	1889104	0
49053	7:00	1429	1	1	360	22.5	−7.5	1	2025601	0
48330	7:00	1429	1	1	127.5	30	0	1	1997501	0
48329	7:00	410	1	1	22.5	30	0	1	1997502	0
46755	7:00	1429	1	1	330	30	0	1	1938601	0
48595	7:00	201	0	2	307.5	37.5	0	1	2008902	0
43952	7:00	201	0	2	180	15	0	1	1810802	0
43854	7:00	1429	1	1	90	30	0	1	1805501	0

Figure 6.19. Trip snapshot result for the 10th iteration of microsimulation for the study area.

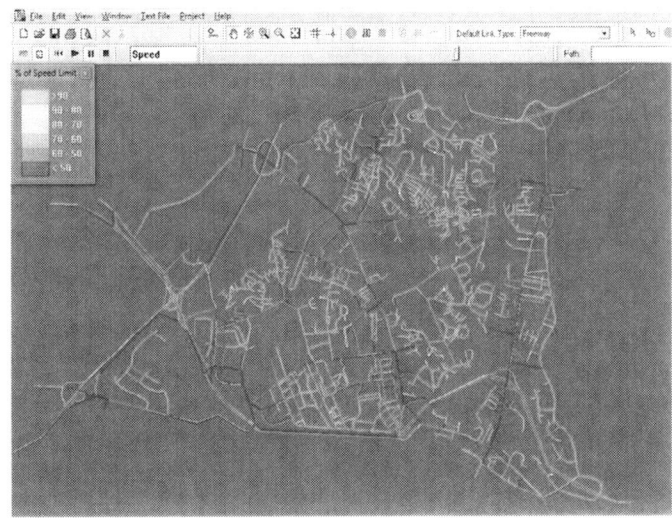

Figure 6.20. Snapshot of speed in the study area using NEXTA at 5:00 PM.

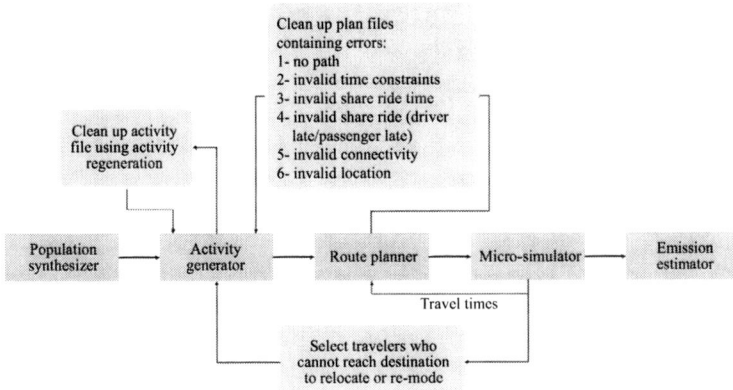

Figure 6.21. TRANSIMS feedback process.

BIBLIOGRAPHY

Baltimore Metropolitan Council 2007. Baltimore Region Travel Demand Model, Version 3.3.

Beckman, R.J., K.A. Baggerly, and M.D. McKay. 1996. "Creating Synthetic Baseline Populations." *Transportation Research A* 30, no. 6, pp. 415–429.

Breiman, L., J.H. Friedman, R.A. Olshen, and C.J. Stone. 1984. *Classification and Regression Trees*. London, UK: Chapman and Hall.

Dijkstra, E.W. 1959. "A Note on Two Problems in Connection with Graphs." *Numerische Mathematik* 1, pp. 269–271. doi:10.1007/BF01386390

FDOT (Florida Department of Transportation). 2014. Traffic Analysis Handbook, A Reference for Planning and Operations.

Hofmann, M. 2005. On the Complexity of Parameter Calibration in Simulation Models. *JDMS* 2, no. 4, pp. 217–226.

Hu, B., J. Shao, and M. Palta. 2006. "Pseudo-R^2 in Logistic Regression Model." *Statistica Sinica* 16, no. 3, pp. 847–860.

Ismart, D. 1990. *Calibration and Adjustment of System Planning Models.* Washington, DC: U.S. Department of Transportation, Federal Highway Administration Publication FHWA-ED-90-015.

Jeihani, M., A. Ardeshiri, O. Ighodaro, and G. Mazloomdoost. 2011. "Cumulative Impact of Developments on the Surrounding Roadways' Traffic." *Maryland State Highway Administration and National Transportation Center at Morgan State University.* Report # SP009B4R, http://morgan.edu/Documents/ACADEMICS/CENTERS/NTC/Cumulative_Jeihani_1112.pdf (accessed April 2016).

Jeihani, M., H.D. Sherali, and A.G. Hobeika. 2006. "Computing Dynamic User Equilibrium for Large-scale Transportation Networks." *Transportation* 33, no. 6, pp. 598–604.

Nagelkerke, N.J.D. 1991. "A Note on a General Definition of the Coefficient of Determination." *Biometrika* 78, no. 3, pp. 691–692.

Park, B., and J. Won. 2006. Microscopic Simulation Model Calibration and Validation Handbook. FHWA/VTRC 07-CR6.

Traffic Operations and Safety Analysis Manual (TOSAM). 2015. Ver. 1.0. VDOT.

TransCAD, Caliper Corporation. http://caliper.com/tcovu.htm (accessed December 2015).

University of South Florida, State Averages for Private Vehicle Occupancy, Carpool Size and Vehicles per 100 Workers, http://nctr.usf.edu/clearinghouse/censusavo.htm (accessed May 2016).

Virginia Transportation Research Council (VTRC). 2006. Microscopic Simulation Model Calibration and Validation Handbook, Report No. FHWA/VTRC 07-CR6.

Zhu, X. 2010. *NEXTA for TRANSIMS.* http://civil.utah.edu/~zhou/NEXTA_for_TRANSIMS.html (accessed February 2011).

CHAPTER 7

CONCLUSION

Transportation problems range from as small as an intersection to as large as statewide or regional problems. At the same time, the focus can be on short-term, medium-term, or long-term plans, where there is usually a correlation between the study area size and study's time horizon. To analyze transportation-related problems, transportation engineers and planners tend to develop mathematical models and simulation tools to rescale real-world problems into sensible, simplified sketches. The notion of simplified should not undermine the sophistication and intricacy level of modeling techniques developed thus far; however, this reminds us of the difficulty of portraying human behavior complexities in transportation models. Travelers—in the form of pedestrians, drivers, cyclists, or passengers—are the key role-players in transportation science. Predicting their behavior, given their various options, knowledge, information, travel modes accessibility, and their socio-economic factors and value-of-time, yet by time-of-day, is still a challenging task for transportation modelers to fully accomplish. Given the variety of transportation problems, their domain and analysis level, it is necessary to have various modeling tools available for different purposes and problem scales, including macroscopic and microscopic simulation, forecasting, and planning tools.

The first step to approach a transportation problem is to define a study area on which to focus. The area may be an intersection, a corridor, a freeway system, or a network of multiple road classes containing at-grade or grade-separated intersections. A real transportation network is replicated into a transportation model to better examine the existing operations and more precisely, investigate the impact of any change to demand, supply, or both in the future. A transportation network model gives decision-makers the opportunity to seek, evaluate, and compare various feasible solutions to address a recurring or non-recurring problem.

Transportation models consist of some key components to constitute a transportation network that are discussed in Chapter 1. Among them, links and nodes are of the most importance, representing streets and junctions, respectively. When coding a transportation network, close attention should be spent on properly coding the number of lanes, direction, lane width, pocket lanes, grade, travel speed, turning speeds (right, left, and U-turns), right of way and stop signs, and so on. Nodes represent intersections, merging and diverging points, and trip generation and transshipment points. Wherever a roadway cross section changes, a new node must be inserted to connect upstream and downstream links. Improper coding of these elements can seriously impact the model's performance and misrepresent the flow of traffic. These elements are among the first checks in calibration efforts. Network zoning is a necessary strategy in aggregating large-scale networks to group trips with common features, such as origin–destination (OD), trip purpose, and travel mode.

Travel demand is coded in transportation models in various formats. Macrosimulation tools require actual turning movement counts and peak hour factors at all intersections. When coding multiple intersections, special attention should be given to corridors' volume balancing between any pair of upstream or downstream nodes. Traffic sources and sinks located in the mid-block that have not been coded in the model can explain some volume imbalances. Examples include parking garages, driveways and alleys, and so on, with trip production or attraction patterns that typically reverse from AM to PM peak hours. Microsimulation tools typically require an OD flow of traffic from a network's entering points to a network's end or mid-points. Improper routings or imbalanced volumes may cause unrealistic turning volumes at intersections. Most of the planning tools and travel demand forecasting models generate traffic at some real or dummy nodes in the center of traffic zones. These dummy nodes that, in fact, do not exist in the real world are known as centroids. They are in charge of pumping zonal traffic into local, distributor, or minor arterial streets. Special care should be given to coding centroid connectors, their capacity, speed, traffic mix, and time-of-day volume.

It is necessary to define some arbitrary points in the network to compare model-generated peak hour flow against field-collected volumes. If model outputs do not fit in an acceptable error range, calibration steps are required, such as adjusting OD matrix, capacity, and saturation flow. Other network checks include lane restrictions, prohibited movements, school zone hours, and railroad gate timings if applicable, and so on.

Traffic control systems are another key component of transportation models that must replicate real-world operations. Signal timing, including green, yellow, and all-red times for each phase; offsets; phase sequence; actuation and detection settings; overlap phases; and pedestrian timings must be coded properly in the model. Calibration measures can be intersection throughput, queue length, and intersection delay time that should be compared against field-measured values.

Travel time or travel speed is a major measure of effectiveness (MOE) in the calibration transportation networks. Collect travel time data in the field is expected, either through multiple travel time runs or through commercial real-time traffic data providers. Under any method, it is critical to obtain an average of multiple observations to avoid discrepancy in travel patterns. When a model's outputs are not within the desired error range, the calibration effort includes adjusting link traveling speed, turning speeds, speed distributions, acceleration and deceleration functions, stop signs, number of turning lanes, and so on.

Collecting model MOEs in stochastic models requires multiple simulation runs to capture an average representation of the network state, which can replicate typical weekday traffic. Calibration is an iterative process that moves the model's output closer to field-observed values of the target MOEs. It is common to notice improvement in one MOE (such as travel speed), but deterioration in another (such as maximum queue length) as a result of a calibration strategy, that is, lane configuration adjustment. Because a combination of multiple calibration techniques may lead to unintuitive results, it is recommended to practice any calibration strategy independent of others to avoid confounding results, and gradually meet calibration thresholds.

Four-step models have been widely used for travel demand modeling. They are easy to use and the required input data is not overwhelming. However, they have various shortcomings and are not appropriate for the current applications, as discussed in Chapter 4. Agent-based and activity-based models were introduced to address four-step model problems. They have been slowly adopted by transportation agencies. These models require extensive amounts of input data, which are difficult to provide.

Travel demand models consist of several consecutive modules or steps, each of which is fed by the previous module. Error in one module would be magnified by affecting the successor modules. Therefore, calibrating and error checking of each module is an important part in developing a robust model. Each module should be calibrated, and the results should represent the real world. The travel demand model should represent the base-year traffic condition, meaning, traffic volumes of roadways estimated by the model should be almost the same as the ones obtained by traffic counts. This comparison cannot be performed for each single roadway link; rather, it is done for total volume on each screen line, cut line, or cordon. A well-calibrated model then should be validated by using for another year for which the results are already known. The model assumptions and parameters need to be revised if the model is not validated.

INDEX

OTHER TITLES IN OUR TRANSPORTATION ENGINEERING COLLECTION

Bryan Katz, *Editor*

High Speed Rail Planning, Policy, and Engineering,
Volume I: Overview of Development and Engineering Requirements
by Terry L. Koglin

High Speed Rail Planning, Policy, and Engineering,
Volume II: Realizing Plans—Obstacles and Solutions
by Terry L. Koglin

Roadway Safety: Identifying Needs and Implementing Countermeasures
by Brian Chandler

Emerging Trends in Transportation Planning
by Andy Boenau

High Speed Rail Planning, Policy, and Engineering,
Volume III: System Operations by Terry L. Koglin

Momentum Press is one of the leading book publishers in the field of engineering, mathematics, health, and applied sciences. Momentum Press offers over 30 collections, including Aerospace, Biomedical, Civil, Environmental, Nanomaterials, Geotechnical, and many others.

Momentum Press is actively seeking collection editors as well as authors. For more information about becoming an MP author or collection editor, please visit http://www.momentumpress.net/contact

Announcing Digital Content Crafted by Librarians

Momentum Press offers digital content as authoritative treatments of advanced engineering topics by leaders in their field. Hosted on ebrary, MP provides practitioners, researchers, faculty, and students in engineering, science, and industry with innovative electronic content in sensors and controls engineering, advanced energy engineering, manufacturing, and materials science.

Momentum Press offers library-friendly terms:

- perpetual access for a one-time fee
- no subscriptions or access fees required
- unlimited concurrent usage permitted
- downloadable PDFs provided
- free MARC records included
- free trials

The **Momentum Press** digital library is very affordable, with no obligation to buy in future years.

For more information, please visit **www.momentumpress.net/library** or to set up a trial in the US, please contact **mpsales@globalepress.com**.

CPSIA information can be obtained
at www.ICGtesting.com
Printed in the USA
LVOW10s2334050218
565450LV00009B/107/P

9 781606 508930